WHAT THE HELL IS SCIENCE?

How to Understand Science, Research, and Scientists, for Adults Who Remember Nothing from School, Are Job-Seekers, Media Creators, Parents or Spouses of Crazy Researchers, and You!

Dr.M

Published by

Dr.M Enterprises

Copyright © 2017 by Dr.M

All rights reserved.

ISBN: 1978344996

ISBN-13: 978-1978344990

Library of Congress Control Number: 2017917395

CreateSpace Independent Publishing Platform, North Charleston, SC, U.S.A.

CONTENTS

Preface   v

Introduction   vi

Chapter 1:   What Is Science? What Is Scientific Research? How in the World Do We Know What Is Really True?

    1

Chapter 2:   How Are Scientists Made? Where Do They Work? What Do They Do? What Are Experiments For? Do Scientists Ever Have Any Fun?

    11

Chapter 3:   What Do Inventors, Engineers, Technicians, Grad Students, and Postdoctoral Fellows Do for Research Projects?

    23

Chapter 4:   How Does Research in Industries Differ from That in Universities? What Is Progress in Industrial Research?

    31

Chapter 5:   Why Is Science So Very Expensive? Who Pays for All the Scientific Research in the United States?

    37

Chapter 6: What Does Grouping Do for Research Projects? What Is Little Science versus Big Science?

45

Chapter 7: On the Nobel Prize and Other Big Honors in Science!

53

Chapter 8: Let's Meet Some Real Scientists and Inventors!

61

Chapter 9: Could Any Modern Research Scientists Actually Be Criminals?

71

Chapter 10: How Can You Find Out What's New in Modern Scientific Research and Technology?

79

Chapter 11: What If You Want to Do Some Volunteer Work with Scientists on a Research Project?

89

Chapter 12: What Can Science and Research Do for You?

95

Index 103

PREFACE

I've been waiting here to be your guide...

So come...

> —Paul Stanley of KISS, singing their song, "Psycho Circus"

    This new book is just for all the millions of adults who know nothing about science! Many need to understand much more about science and research because they are recently married to a scientist, have good friends doing research, work in a science-related job, or have children needing help with science classes in school. Here you will learn to understand the basis for science, research, and scientists in today's world.

My goal is to guide you to understand much more about what science and research are for and what scientists actually do! This is not a textbook, so it has nothing to be memorized, equations to be copied, or exams to be taken!

Most adults nowadays feel totally separated from science and have never talked to a real living scientist. Their view of researchers is shaped by the Hollywood and television caricatures showing scientists as weirdly funny people in white lab coats who create monsters or make mutant diseases. Bunk! Those entertainments are grossly distorted from reality! By peeking at the lives and personalities of some real scientists and inventors who I will introduce you to, you will see that they are actually just fellow people!

Dr.M (Dec. 8, 2017)

INTRODUCTION

This new volume will not make you an expert on science, because it is not a textbook! Don't worry about memorizing equations, doing homework or taking exams, because there are none here! Instead of teaching you science facts, it will greatly enlarge your *understanding about science*; if you want to learn facts about science, then you should study a good textbook. Instead of teaching you how to conduct research, it will enable you to *understand how research investigations proceed*. Instead of teaching you how to be a scientist, it will let you *understand what real scientists work on*.

In other words, you will gain a whole big bunch of new understanding from this book! Your expanded understanding will let you stop being afraid of research, better grasp what is new or controversial in science, and see how scientific research actually influences you every day.

This new book is designed to be fast and easy to read for everyone no matter what their background, and useful to all adults trying to comprehend science, research, and scientists. You can readily work at your own pace. As a bonus, it also will give you a much more interesting life!

Science is strongly related to all of us since it literally involves everything we do. Knowing how to seek the knowledge coming from science and research on the Internet makes it easier to find good solutions for dealing with life's everyday small problems; this book shows you how to do that!

Who is Dr.M, your guide in this new book?

I was most happily educated in the college, medical school, and graduate school of The University of Chicago. My professors trained me to always work very hard, to freely ask many questions about anything and everything, and to try to be clearly aware of alternative viewpoints. My PhD in cell biology was followed by two postdoctoral appointments; the first was in France and the second was in the United States. During my thirty-five years as a faculty scientist, I have taught lecture courses, laboratory course sessions, and interactive discussions for undergraduates, graduate students, medical students, and clinical residents. My grant-supported research experiments have been conducted as a biophysicist, cell and molecular biologist, and structural biologist. I love doing laboratory research!

CHAPTER 1

What Is Science? What Is Scientific Research? How in the World Do We Know What Is Really True?

If you don't know anything at all about science but are curious why your doctor just put you on a special diet, then this book is for you! Science classes in your youthful schools were boring and featured memorization, so you and many in the United States can now recall nothing! People love to laugh at the Hollywood portrayals of scientists as weird creatures from some other planet; that popular fantasy leads to the very mistaken belief that science and research are irrelevant for daily life. Yet what your physician did to care for you is directly based upon medical research!

With this book, you will learn about science without needing to memorize any equations or learn hundreds of special terms! Also, by learning to understand the basics of science, your life will become much more interesting! After seeing how scientists are trained to do research work and how scientists and engineers develop new knowledge and technology into commercial products, you will start to realize that you no longer need to be afraid, confused, or distressed by science and research!

To begin to understand how we know what is really true, this first chapter provides you with a very fundamental background about science and research. The following chapter will build upon this foundation by explaining how scientists are made, what they actually work on, and how they have fun.

What exactly is science?

Science seeks to find evidence for exactly what is true! We all have been taught many "facts" in school, but how is it known what is really true? Just because a teacher, authority, or group says something is true does not make it so! Scientists carefully conduct detailed research studies to gather evidence about truthfulness.

FIGURE 1.1: *Although modern kids know very much about personal computers and spending money, they have an utterly false image of science, and so do most of their parents!*

Truthfulness and falsity are based upon the evidence coming from organized investigations by scientists and other scholars. The truthfulness of any supposed fact is always open to further study. *Scientists conduct research studies to gain knowledge that is new and true.* However, the clear truth coming from science can unfortunately become changed, confused, or denied by nonscientists from businesses, government, or politics.

Knowing the truth is essential. Is eating cholesterol really bad for our health, or could it be natural and good? Is Earth really warming too much, or is that just normal variation? Are some common weed killers harmless to ingest, or do they actually poison all of us? Unlike falsity, the truth makes valid predictions, contains no internal conflicts, and is consistent with knowledge from other areas of study. Once a new fact is established, it remains open to further testing by additional research.

What are the major parts of science?

Science is divided into three main branches according to what is being studied. These are: 1) biology and medicine, 2) chemistry, and 3) physics. Each branch has subdivisions. **Biology/medicine** includes cell biology, field biology, genetics, immunology, microbiology, molecular biology, neuroscience, pathology, physiology, plant science, and so on. **Chemistry** has analytical chemistry, biochemistry, chemical engineering, inorganic chemistry, nanochemistry, organic chemistry, polymer science, and so on. **Physics** has astronomy, fluidics, materials science, nuclear physics, quantum mechanics, theoretical physics, vacuum science, and so on. Most of these subdivisions are broken down further into yet smaller divisions.

FIGURE 1.2: *Many people are so ignorant that they still believe most scientists are mad maniacs who create monsters in their laboratories!*

What is scientific research?

Anything and everything is open to question by scientific research! Research experiments in any branch of science systematically study the activities, composition, dynamics, genesis, interactions, properties, and structure of matter in various natural or synthetic specimens. **Experiments** are an organized way to collect that information (i.e., the **data**). Often, experiments make a controlled change to determine cause and effect relationships. Research experiments can be done either in laboratories or in the field; data can consist of measurements or observations. **Laboratories** are dedicated rooms filled with instruments and special supplies so scientists can conduct research experiments to gather new data.

Experiments are conducted to try to answer **research questions** (e.g., "Is Earth actually warming?" "Do other stars have planets?" "Why can't humans regenerate their fingers?"). Measurements in any **assay** (e.g., weighing something or measuring the temperature) are done in a standardized way, termed a **protocol**. After collecting many results, all the observations, measurements, and experimental findings obtained (i.e., the **research data**) must be analyzed to derive solid conclusions about what is going on; ideally, these results and conclusions should answer the research questions posed (e.g., "What is the cause of some observed effect?").

What are the main kinds of research?

Scientific research has several operational divisions. For **basic science** and **basic research**, new knowledge is sought for its own sake without any direct regard to possible uses or commercial value. Basic research asks questions like, "What is the role of iron in ocean water," or, "What is the structure and composition of membranes in bacteria growing in hot springs versus those growing in Hudson Bay?" Many questions in basic research often seem very esoteric to nonscientists but are actually meaningful.

Basic science differs from **applied science** and **applied research**, where experiments have the goal of developing useful activities or properties with some material object, process, or product. The general goal of applied research often is to change or improve something that already is known from basic research; basic research studies are the precursor to all applied research projects.

With additional input by engineers (**engineering research and development**, often abbreviated as **R&D**), the commercial output of new, improved, or modified products can be completed. Applied research is in the middle of a progression going from basic science to applied science and on to engineering development; this leads to commercial output.

A different way to divide research into categories is based on where the studies are conducted (i.e., within a **laboratory** [lab research], in nature [field research], at hospitals [clinical research], in outer space [space research], at planetary poles [Arctic or Antarctic research], and so on). Additionally, research can also be subdivided based on which special **instrument** (a microscope, telescope, and so on) or **method** (e.g., chromatography, cytochemistry, spectroscopy, and so on) is utilized to gather data. These categories of research are simply different approaches to gather evidence for what is really true.

What is a research project?

All the different investigations in any scientific study constitute one **research project**. Each project uses experiments to try to answer important research questions. The results of a research project or a group of related studies are presented to the world by publication of reports in science journals and by new patents.

A typical research project is quite analogous to constructing a new building! One leader supervises all the many different aspects of work in both construction projects and research projects. Specialization of construction workers is traditional (e.g., architects, carpenters, electricians, ironworkers, masons, painters, plumbers, and so on); this corresponds to different technical specializations of the staff working in a research lab. Although everything is carefully planned in advance, it is not unusual that some changes must be made in both construction and research projects while the work is underway. When a new building is finally finished, photographs are taken for advertising, publicity brochures, and press releases; these documentations correspond nicely to the research reports published by scientists in science journals.

What are professional scientists, and what do they do?

First and foremost, **scientists** are people who conduct research investigations. They typically have a great curiosity and a personal drive to discover new knowledge. Scientists conduct research **experiments** to try to answer **research questions**. For a new study, the first questions examined are usually quite general (e.g., "What is its size?" "What is its structure?" "What is it made of?" "How was it formed?" "What are its functional properties?" "How does it work?" "What does it interact with?"). Then, more detailed specific questions can be posed (e.g., "What is its atomic structure?" "How does its composition vary?" "Where and when does it appear?" and "What happens when temperature is changed?"). Scientists get answers to these research questions by measuring and probing, carefully analyzing all the gathered data, and then generating valid conclusions.

Some research scientists study an entire forest or galaxy, while others direct their investigations to specific species of trees or types of stars. With the diverse universe of natural and synthetic research subjects, it is not surprising to find that almost all professional investigators today are **highly specialized**; each conducts research only on one or a few **subjects** and **specimens** (e.g., liquid plastics, living sperm cells, magnetic nanostructures, metal catalysts, noble gases, plant viruses, protein assemblies in biomembranes, and so on). In addition to all the many scientists studying real specimens and physical materials, some other scientists work at the level of untouchable objects and theories. Much more about all scientists will be presented in the next chapter!

How do we know what is really true?

Finding new truth is the goal of scientists, but how is it decided what is true and what is false? For any new or old truth, the number one question for scientists always is, **"What is the evidence?"** Evidence for science is the research data; the better the evidence, the stronger the conclusion that something is really true or must be false. If a flock of scientists cannot reproduce some research result published by another scientist, then that cannot yet be accepted as proven. See Figure 1.3 for some very noteworthy quotes about **The Truth**!

> **Not being known doesn't stop the truth from being true: RICHARD BACH!**
> (http://art-quotes.com/getquotes.php?catid=317#V0CEK_Brj-e)
>
> **Beyond a doubt truth bears the same relation to falsehood as light to darkness: LEONARDO DA VINCI!**
> (http://www.brainquote.com/quotes/keywords/falsehood.html)
>
> **The truth ... is that a billion falsehoods told a billion times by a billion people are still false: TRAVIS WALTON!**
> (http://www.searchquotes.com/search/False_Truth/)

FIGURE 1.3: Noteworthy quotes about The Truth!

Most of us accept the common views about what is true by virtue of authority, tradition, and personal experience. However, anything can be questioned by research scientists. The validity of any new research finding can be contested by other scientists who have acquired differing experimental results; the ongoing controversy between scientists about whether vaccines and vaccinations can have some bad effects is a great current example of a dispute about what is really true. If something widely accepted as being true is shown by new research results to actually be wrong, that still represents good and useful progress for science.

Is scientific research ever guaranteed? No!

All working scientists know that research experiments often do *not* work exactly as planned! In addition, mistakes by scientists and lab workers can and do occur. These two facts mean that *there never are any guarantees in research*! The most frustrating of all difficulties in doing lab studies occurs when a spectacular and hoped-for result is obtained on the very first conduct of an assay, but then this result cannot be repeated despite many attempts—help! Experimental results that cannot be duplicated at least one more time do not officially exist and cannot be reported or published. Research scientists necessarily need patience and flexibility to successfully conduct research, and they must become experts with problem solving. Of course, having good luck certainly also helps!

If there are no guarantees in research, how do scientists make progress by conducting experiments in a lab? Most experiments do have expected results based upon knowledge about what previous or published experiments showed. Scientists must use that awareness to guess at why an experiment gave results different from their expectations and then either modify the experiment before repeating it or change the expectations. Such debates with oneself are inherent in any research project and are actually good; as professional scientists gain more and more experience, they become skilled at dealing with this operational problem.

Essentials to be learned in chapter 1!

1. Scientific research provides data about what is true; the most important question asked by scientists is, "What is the evidence?"

2. Anything and everything can be questioned and studied by scientists conducting research investigations; new research results can verify, deny, or change any supposed truth.

3. Basic research is the basis for all later applied research; commercial production depends upon a sequence of efforts with basic research, applied research, and engineering development.

FIGURE 1.4: *All children today know with certainty that something absolutely must be true if it's ever shown on TV! Lord help us!*

CHAPTER 2

How Are Scientists Made? Where Do They Work? What Do They Do? What Are Experiments For? Do Scientists Ever Have Any Fun?

Today, there are more scientists working on research than at any other time in the history of our world! Research scientists frequently have an enormous curiosity, so they ask very many questions about everything. They have a personal determination to discover new and true knowledge through their research investigations. Most scientists holding a doctoral degree conduct their research studies in **academia** (universities, medical schools, research institutes, government research centers) or in industries (small businesses, big commercial companies, very large global corporations), but some now work elsewhere completely outside of laboratory efforts.

Few adults have ever talked to a scientist, whether on a commuter train, in a tavern, at an exercise center, or at a party. Even fewer have ever visited a research laboratory and seen with their own eyes what goes on there; I will not reveal any embarrassing secrets here! These conditions strengthen the very common belief that scientists must be crazy and therefore research has zero importance for anyone's daily life. We will soon see that this very common belief is terribly mistaken!

In this chapter, you will learn how scientists are trained and what they actually do as professional research workers in universities or industries. To aid readers in getting to know and understand some real research scientists, you will meet several later in chapter 8, and you'll get a peek at their personalities and adventures in science.

FIGURE 2.1: Everyone has their own definition of what is fun! For doctoral scientists, making new discoveries and research breakthroughs, completing a difficult study, publishing results or getting patents approved, and success in receiving research grant awards, all are great fun!

How are professional scientists educated and trained?

To conduct experiments that answer research questions requires many years of education and practical training. After obtaining a bachelor's degree in college, these students spend three to ten more years in a **graduate school**. During their first year, **graduate students** will rotate through the research labs of prospective faculty mentors; practical instruction for researching begins during these rotations via observing different experiments in progress and learning how to operate some research instruments by 1:1 instruction. These young scientists soon learn to use their eyes and hands as well as their minds.

After picking a mentor to be their **thesis adviser** supervising the **thesis research project**, graduate students develop a thesis proposal including important research questions and doable experiments. Then (finally!), they begin to conduct hands-on research investigations. Usually, grad students work on their research studies for at least three years before they are able to defend their **doctoral thesis** (i.e., dissertation) in front of a board of faculty scientists; success in the **thesis defense** means that they will be awarded a doctor of philosophy degree (**PhD**) in their chosen branch of science.

Graduate school education almost always results in acquiring expertise within only a small part of some larger branch of science, developing expert skills in operating only a few different research instruments, and having the ability to obtain good data using only a few approaches for experiments. Therefore, new PhDs usually then become **postdoctoral research associates** or **fellows** in the lab of a different professional scientist. Postdocs are research trainees who receive advanced practical experience in conducting experimental investigations; they work with increased self-direction, are exposed to additional research topics, master more kinds of research instrumentation, and learn how to critically evaluate their own research findings and those of other scientists. Thus, postdocs greatly deepen their expertise beyond that of newly graduated PhDs.

Typically, several solid publications in science journals are produced during the postdoctoral period. Postdoctoral positions usually are renewable for several years. When good jobs become hard to find, young professional scientists will often accept several consecutive appointments as a postdoctoral researcher in the same or several different labs.

Where do doctoral scientists work?

After all that time, eager postdocs are finally ready to find a professional job! A PhD in science provides many opportunities for employment. Scientists have traditionally found jobs in universities, industries, hospital research labs, or government science centers. New scientists in jobs at industrial research and development (**R&D**) centers undergo much the same graduate education as their colleagues aiming to be faculty scientists; however, some industrial positions skip the years of postdoctoral training, since new hires will receive extensive specialized instruction and supervised practical experience on the job.

Today, new doctoral scientists can also increasingly find good employment opportunities outside of academic or industrial labs. They work as professionals with jobs in advertising, computation, finances, media, technology, and so on. Some now even start a brand-new small business!

What do professional scientists do in their daily work activities?

The fundamental aim of scientists is to discover knowledge that is new and true. Faculty scientists in universities primarily work on researching in labs and teaching in classrooms. Other activities include writing manuscripts for research reports to be published in science journals, reviewing manuscripts from other scientists as an invited critic and referee, composing applications for research grants, serving on local or national committees and review boards, mentoring graduate students and postdocs, and participating in national and international science meetings. They are always very busy!

The following listings reflect my own experiences as a biomedical faculty scientist employed at several academic institutions in the United States. Details will be somewhat different for research scientists in other disciplines (e.g., astronomers, chemical engineers, clinical researchers, computer engineers, field workers in geology or biology, and so on), and for the numerous investigators working within industrial laboratories (i.e., industrial scientists conducting applied research and development).

Common **daily activities** for faculty scientists in universities include:

1. thinking (e.g., wishing, wondering, and worrying);
2. planning;
3. discussing;
4. supervising lab staff;
5. writing and revising applications, manuscripts, reports, and so on;
6. solving problems;
7. specimen procurement and preparation;
8. hands-on research operations to collect new data;
9. conducting computerized analysis and display of acquired research data;
10. forming computer displays of the processed data;
11. educational activities for teaching and testing;
12. equipment maintenance and repair;
13. attending conferences and meetings;
14. processing e-mail, postal mail, and telephone calls;
15. reading;
16. adjunctive activities such as setting up new protocols for use by lab staff, composing new software, developing new methodology, meetings with administrators, collaborators, engineers, and so on; and,
17. drinking many caffeinated beverages!

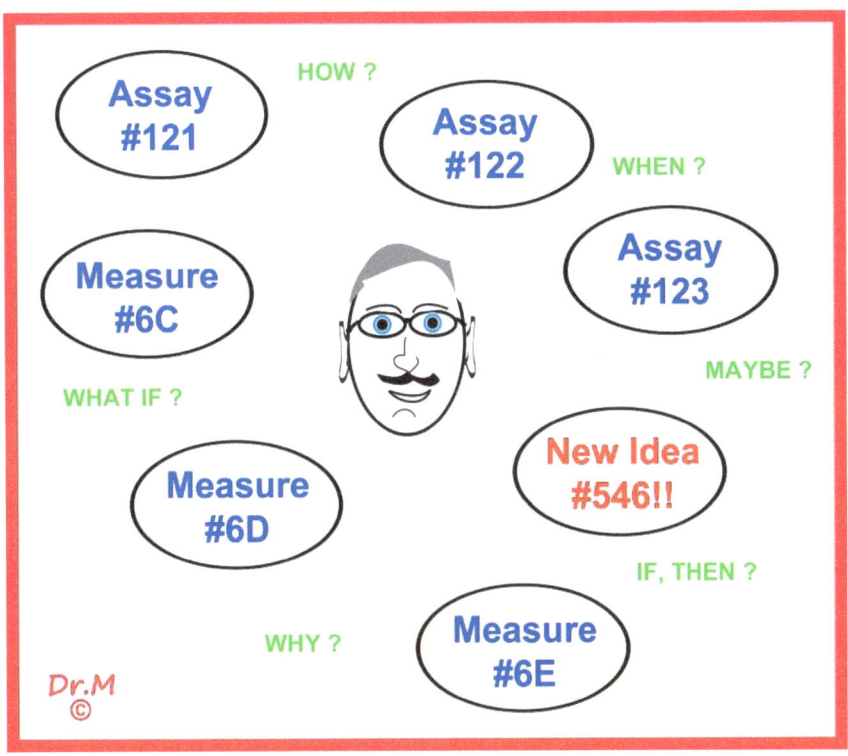

FIGURE 2.2: Research scientists just love to ask questions! As data are being gathered, dozens of new questions often arise!

What is a typical workday for young faculty scientists at a US university?

Based upon my own experiences and observations, the following are quite typical activities during any workday for young faculty scientists in universities. For researchers working in industrial labs, large hospitals, or government research centers, there will be some differences, but many daily activities are quite similar.

MORNINGS:
1. Travel to work, have morning coffee, tea, or Diet Coke, and check e-mail.
2. Briefly review planned schedule for today's activities.
3. Write for publications, reports, and research grant applications.
4. Prepare for today's educational activities in the classroom or teaching lab.
5. Have brief discussions with lab personnel about yesterday's results and plans for today's work.
6. Engage in other personal activities for either teaching or research.

AFTERNOONS:
7. Eat lunch.
8. Go to committee, departmental, or other meetings.
9. Review ongoing experimental results with postdocs.
10. Conduct research bench work and other hands-on activities in the lab.
11. Complete data analysis, deal with problems in experiments or instrument operation, plan scheduling for the future, and so on.
12. Take care of telephone calls, e-mails, and postal mail.

EVENINGS (in the faculty office or at home):
13. Engage in reading activities.
14. Do other work activities (e.g., data analysis and display, looking up information, revising manuscripts, writing documents, and so on).

15. Travel back home to eat, wash, pull up the sheets, and at last mercifully fall asleep.

The total time for this typical schedule far exceeds the standard US workweek of around thirty-five to forty hours. All professional researchers also need some time for family life, social events, illnesses, annual vacation, and participation in church, schools, and other organizations. This heavy schedule makes it obvious that there will be only limited time for new faculty scientists to conduct hands-on research experiments in their own lab, even though that is exactly what their very lengthy education and training was for!

How are scientific research studies conducted?

Each research investigation uses a number of sequential steps:

1. selection of a topic or subject for study;
2. selection of specific research questions to be investigated;
3. design of the planned experiments to be conducted;
4. preparation of materials, specimens, and instrumentation needed for the study;
5. running the experiment and recording the results (i.e., collecting data);
6. analysis, evaluation, and statistics of the collected data;
7. interpretation of these new results (i.e., "What do they signify?");
8. comparison of the new results to other acquired or published data;
9. drawing conclusions regarding the research question;

10. repeating an experiment to gather additional data;

11. formulating new research questions; and,

12. planning and running the next set of experiments.

This entire process sequentially cycles around and around as the several planned studies within one **research project** proceed. Often experiments must be restarted or repeated. Each basic study focuses on a well-defined question or separate subject (e.g., "How much gold is dissolved in a tropical ocean?"). Later, applied researchers can work with questions about potential importance for commerce and society (e.g., "Can gold be harvested commercially from ocean water?"). Scientists at universities publish their results and conclusions in a science journal; scientists and engineers at industrial labs use their research findings mainly to submit proposals for new patents.

Carrying out research can be viewed as analogous to composing a piece of music or a poem! The actual course of *research only rarely is a linear chain of experiments*. Usually there are branches and parallel studies (e.g., certain data might involve chemistry, while other data looks at synthesis, structure, or electrical properties); these are similar to different formats for poems. All the new data must be analyzed together in order to form valid conclusions; that complex set of activities is similar to the interplay of melodies in a song. The end results of research, music, and poetry usually benefit science, society, and commerce.

Do scientists ever have any fun? Yes, indeed!

Scientific research certainly is serious work, but it also can be fun! The thrill of being the first human to discover something is always exhilarating and usually can make up for all the long hours and many job frustrations. Answering a research question that no one was able to deal with before, definitively proving that a disputed conclusion is true, or finally succeeding in publishing a manuscript after several difficult and prolonged rounds of revision will all make any professional research scientist feel very uplifted. The fun and satisfaction of publishing a new journal article or receiving a renewal of a grant are usually shared with the entire lab staff at a restaurant or home party.

A senior academic scientist who works to train younger graduate students to become proficient in research and then watches them develop and successfully complete their thesis study is certainly happy to give birth to new scientists. Seeing a postdoctoral fellow landing a terrific job is always very satisfying to his or her mentor. Even ordinary scientists just doing their daily lab work enjoy publishing in science journals, getting a patent approved, seeing distant colleagues and friends at the annual science meetings, and being called upon to referee a manuscript for a professional science journal or to mentor a new faculty appointee. Some research scientists, whose interpretations or concepts are initially widely disputed or downplayed by other scientists, only achieve their fun decades later after they finally show by continued difficult research efforts that their very unconventional and controversial concept really is valid and true.

For me, from my undergraduate research in college right up until the present day, it always has been totally thrilling and enormous fun to operate an electron microscope and to look at different specimens. This involves my working in total coordination with the complex instrument as if we are a single entity; for me, it is a magic fusion of man and machine, a convergence of human eyesight and computerized electronics, and a delightful merging of science with art. Recording good images with electron microscopes involves developing critical judgment and can be an act of creativity (i.e., what appears in the image is directly related to how the specimen was prepared and how the instrument is adjusted before recording it). Taking a first look at a new specimen is often an adventure filled with many surprises. In secret, I am convinced that the electron microscope is the very greatest toy that has ever been invented for adults! The 2017 Nobel Prizes in Chemistry were awarded to three research scientists developing and working with a very special type of electron microscopy!

Essentials to be learned in chapter 2!

1. Scientists have the fundamental aim of finding new knowledge that is true; they do that in order to try to answer research questions.

2. Education and training for doctoral scientists typically takes many years!

3. Professional scientists working on research investigations have traditionally found employment in academia and industries; today, some also are employed completely outside the laboratory, and others start their own new small business.

4. The busy schedule of daily work activities for faculty scientists means that they are often very short of time for their family life and for other personal activities or interests.

5. Scientists working hard on research do indeed have some fun!

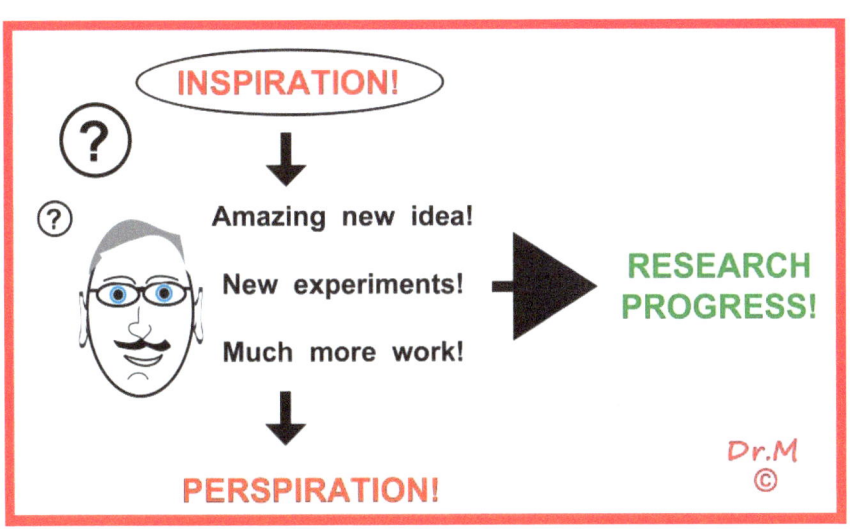

***FIGURE 2.3**: Research progress is the product of both inspiration and perspiration by creative professional scientists! Not depicted here are all the different specialized lab staff working on the research project alongside their boss (see chapter 3 for much more on lab staffing)!*

CHAPTER 8

What Do Inventors, Engineers, Technicians, Grad Students, and Postdoctoral Fellows Do for Research Projects?

Other workers directly help professional scientists conduct research investigations and make new discoveries. In this chapter, we will look at the role of these associated workers who contribute much to the progress of scientific research. Emphasis here is given to research laboratories in universities, medical schools, hospitals, and governmental research centers (i.e., academia), while the next chapter will focus on examining the corresponding operations for investigators working in industrial labs.

What do laboratory bosses contribute to research progress?

The head scientist within any given laboratory or project is termed the **principal investigator (PI)** in universities (i.e., the lab boss and recipient of a research grant), or the **group leader** or **lab chief** in industrial R&D labs. PIs design a research project when they compose an application for a new research grant. These doctoral scientists supervise daily work activities in their lab; provide instruction on research procedures, methodology, and instrumentation; guide data collection and analysis by lab staff; generate and submit manuscripts and patent applications; compose applications and requests for funding new research projects; and deal with local administrators.

In academia, the PI has the essential role of acquiring money for the salaries of all lab staff. Because they always are very busy, many of these bosses in academia shift their identity to become

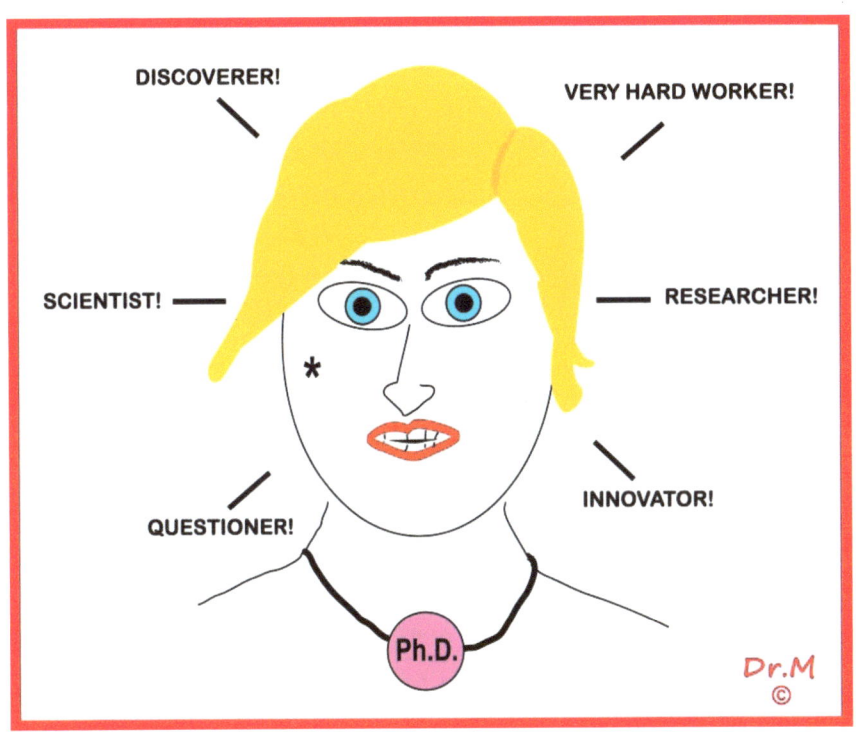

FIGURE 3.1: *Postdoctoral research fellows (and gals!) always work hard and contribute very much to the fun, productivity, and success of their mentor's laboratory!*

research managers who sit at a desk and no longer do any hands-on benchwork in their lab(s). They work on research vicariously and are actually a specialized type of administrator. This situation is very frequent now, but is largely unknown to the general public.

What do inventors contribute to research?

An **inventor** is the first person to define some new idea and then make a working prototype for a new device or process. Typically, inventors figure out how to put their mental idea into practice by actually constructing something new and working out its operational capabilities for practical usage. **Amateur inventors** can be you or anyone; they usually do not have any special advanced education, but they use their creativity and experience to generate a new idea. **Professional inventors** repeatedly produce different new devices or operations and are having some commercial success. Most inventors are not doctoral scientists, but many scientists also are inventors; for example, scientists often need to design and build a new research instrument or modify an existing one so it has wider or more accurate applications for their research.

Most people have heard about such world-renowned professional inventors as Alexander G. Bell (telephone), Thomas A. Edison (motion picture camera and incandescent light bulb), and Edwin H. Land (Polaroid cameras and films). What about the inventors of today? For the story of a truly fantastic modern inventor, see chapter 8 and watch two inspiring Internet videos about the late **Artur Fischer**: "Artur Fischer—Wall Plug, Synchronized Flash, and Many More" (https://www.youtube.com/watch?v=Ke5K4S-Cjj8) and "Artur Fischer in His Own Words—Winner of the European Inventor Award 2014" (https://www.youtube.com/watch?v=hmkZ_ipbY90).

What do engineers contribute to research projects?

Different kinds of **engineers** generally make designs, modifications, or extensions of a previously invented device, object, or process. They usually have a professional education and advanced experience, enabling them to solve practical problems and evaluate the effects of making simple or extensive changes in design and operation. Commercial companies often use both engineers and applied research scientists for their product development. For research in academic labs, engineers are used only infrequently; however, when a new or modified research instrument is needed, consultation with engineers becomes vital.

What do graduate students contribute to research projects?

Science students usually spend three to ten years in a graduate school receiving advanced education and personal instruction while they conduct their thesis research. **Graduate students** work for at least several years to get important new research results that fulfill the requirements for a PhD degree. Grad students learn to critically evaluate research data and to ask probing questions. They are indeed useful to their thesis adviser by providing important new research results that advance their lab's productivity. Nowadays, their research work often satisfies some of the specific aims for the research grant of their thesis adviser; alternatively, grad students enable a pilot project to be developed and tested for subsequent use in applying for a new research grant. Some **undergraduate students** can work on lab research for shorter periods of time (e.g., one to twelve months) via special programs at their colleges.

What do laboratory technicians contribute to research projects?

Research technicians assist their supervising scientist by carrying out many routine lab tasks (e.g., ordering research supplies, preparing media and stock solutions, washing and cleaning, and so on), as well as handling some more specialized activities (e.g., computer processing, data collection, sample preparation and analysis, and so on). Techs usually have a bachelor's or master's degree; however, job experience and technical skills count much more than their earlier coursework. Research technicians can be invaluable members of the lab team due to their ongoing experience and practical expertise with many different lab operations. They can vitally assist with making research progress in the lab as a reliable pair of additional hands and eyes; they are encouraged to ask questions and discuss how to activate new experiments in practice.

What do postdoctoral fellows contribute to research projects?

Postdocs work full time on research semi-independently while obtaining additional practical experience with methodology and instrumentation. They expand their technical capabilities and learn to use new research approaches by working with a different mentor than the one supervising their thesis research. Postdocs are workhorses and are often the heart of lab productivity; typically, they produce several good research publications. After completing their postdoctoral appointment(s), they can enter the job market as new professional scientists.

How do collaborators contribute to research projects?

Research collaborators in academia are other professional scientists who provide additional expertise or special specimens for the research project of a fellow scientist. They can be located in the same institution or have their home base at some distant site. Collaborators bring a deeper dimension to a research project; both the visitor and the host scientist benefit from working together, and they typically coauthor publications. For industries, research collaborators work within the same company by visiting other labs in their building or one at a distant site.

On the importance of coworkers and teamwork for conducting good research!

All the personnel described above function as **coworkers**. Each member of any lab, whether this is two or thirty persons, has responsibilities that are very important for success of the entire group. Each worker is trained and specialized, gathers valuable experience with their specific roles, and functions in coordination with the other lab workers.

It is not unusual, when someone must be absent due to illness or vacation, that a coworker takes that person's place, so one individual is temporarily doing the work of two. Although a tech cannot completely fill in for a postdoc, all within any lab should have some understanding about what other lab members are doing. The PI must foster teamwork by keeping everyone well informed on how research in the project is progressing; it is essential that the boss holds occasional group meetings and has some private discussions with individual team members. In the best cases, all of these leads staff coworkers in a good research lab to become the equivalent of a **laboratory family**! The operations of smaller and larger research groups will be examined later in chapter 6.

What determines the degree of success for a research project in modern academia?

Actual laboratory research is centered on a team of interacting coworkers. The different research employees in a lab of any size ideally should all work together as a **research team** with shared responsibilities; this is very true for research labs in both academia and industries. The traditional image of some brilliantly creative scientist working for years all by himself or herself probably doesn't exist anymore in today's world. Efforts by individuals remain very important for research, but progress and success in any branch of modern science demand the use of **teamwork**. Teamwork in small or larger groups is vital to make progress in science and research!

All research labs try to make discoveries that will be considered important by other scientists. **Research breakthroughs** stimulate new directions and new approaches for further investigations; to achieve them is not so easy, but they are highly sought because they benefit the progress of science.

Success in academic research is traditionally measured by such mechanical standards as the number of publications produced, the number of research grants acquired and renewed (see chapter 5), the number of dollars awarded in research grants, and the number of graduate students and postdocs being mentored. More subjective evaluations look at the reputation of the individual among other scientists in the same field, the quality of their research experiments and publications, formation of collaborations with famous scientists, and prominence within a national science society.

Essentials to be learned in chapter 3!

1. Modern labs have different research staff workers conducting a variety of essential operations for different parts of a scientific investigation.

2. The final success and output depends upon the individual contributions of all the different staff working as a dedicated team of coworkers.

3. Cooperation and active dedication to teamwork permits all lab workers to see their research group as nothing less than a laboratory family; the different individuals in this setting all are important for successful completion of the entire research project.

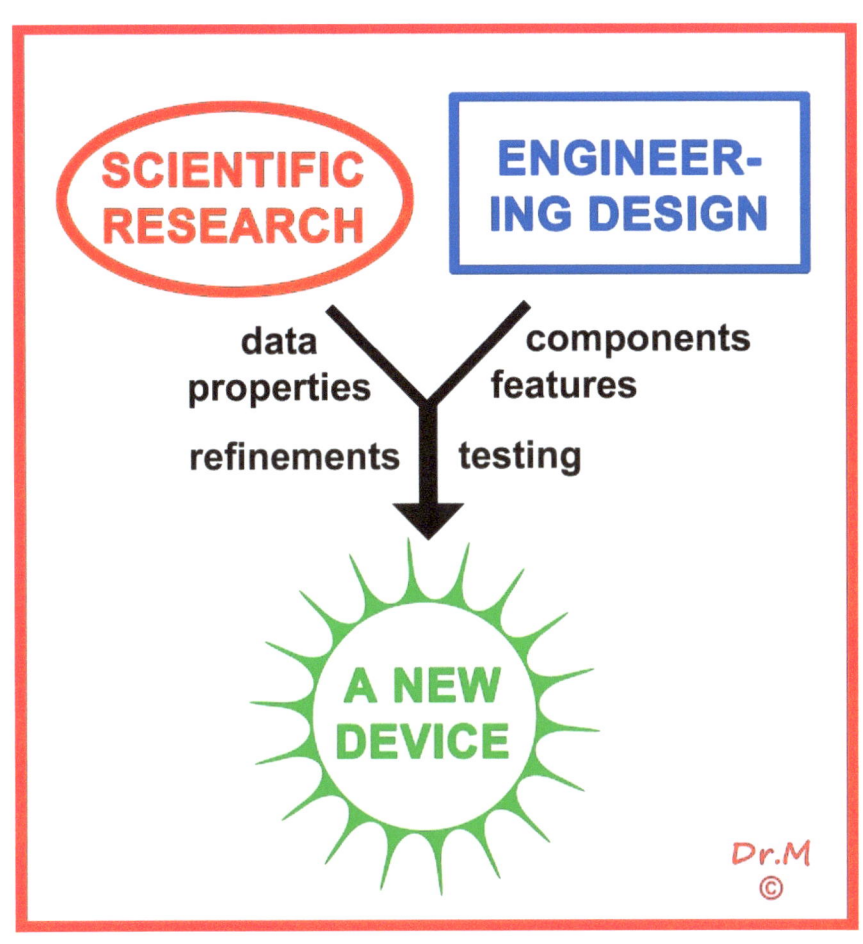

FIGURE 3.2: Basic and applied scientific research interact and combine with engineering design and industrial developments to result in new commercial devices!

CHAPTER 4

How Does Research in Industries Differ from That in Universities? What Is Progress in Industrial Research?

Scientific research finds its traditional home inside laboratories, both at university science departments and at industrial research and development (R&D) centers. While high-quality research investigations at both locations are important and useful for society, their several operational activities differ quite notably. This chapter examines and explains some distinctive features for research operations at these two different settings.

Scientific research in industries and universities is both similar and different!

Walking into a laboratory room within a university science department or an industrial R&D center shows similar general features for both **organization** and **research operations**. At universities, a faculty scientist holding a research grant is the principal investigator (PI) and is in charge of all daily research activities in the lab. Although a group leader or lab chief heads each lab within an industrial R&D center and is in charge of all daily activities, the entire industrial project is usually directed by a highly placed administrator (e.g., project director or research director). While that administrative official interacts with the lab chief(s), only the latter actually works in the lab and corresponds to the PI in academia.

FIGURE 4.1: *Amateur inventors are numerous! New ideas by inventors can be followed by applied research and engineering developments to make new commercial products within industrial R&D centers!*

Money has a very large significance for research everywhere. Funding for research in industrial labs comes from the company's profits; decisions about expenditures are made by lab chiefs after administrators have approved a budgetary allocation. Funding for research in universities comes from research grants awarded to applications by individual faculty scientists; each PI makes decisions about expenditures from their awarded funding. Total funding for all research and development activities inside all US industries spends billions more dollars than do all the federal science agencies issuing grants to support research activities in academia.

Basic research is mainly conducted by faculty scientists in university labs. **Applied research** is centered on staff scientists and engineers working in industrial labs. In academia, faculty scientists doing basic research mostly interact with staff **engineers** only infrequently; applied researchers in industries have frequent interactions with engineers as development progresses and plans for commercial production get finalized.

For **staffing** in industrial research labs, graduate students and postdocs are traditionally fewer, but engineers are more numerous than in university labs. Lab techs are numerous at both locations. **Research equipment** tends to be newer in industrial labs, largely because purchase of the latest instruments is more easily justified as being essential in order to initiate a new and very desirable research project. For labs in academia, the common alternative to buying the newest expensive research equipment is to keep older instruments in operation or make arrangements to use new instruments at some other location.

Industrial research concentrates upon acquiring new **patents** and developing improved commercial products. Patents are a legal license to produce a commercial product free from competition by other companies for some years. University research mostly produces **research reports** published in scholarly science journals.

What about project selection and research freedom in industries or academia?

There are large differences for project selection and funding at the two locations. For industries, the **research question** or **subject** is selected by some company official (e.g., the chief scientist or research director); individual labs and scientists are funded internally by the company's budgetary allocations for research operations needed to advance profits. In academia, the research question or subject for investigation is decided by individual PIs; financial support is made via applications for external funding that succeed in getting awarded a research grant. In other words, determination for project selection by administrators in industry is made by the research grant system for university scientists.

In academia, **research freedom** to choose a subject for study and the research methods to be utilized requires academic scientists to get a research grant. That freedom is largely given up by scientists researching in industries since the subjects for new research are decided by administrators. In principle, an industrial scientist can propose a new research project and submit this for review by the lab chief; if approved, it then goes upstairs for review by higher administrative authorities. Of course, if an industrial scientist finds employment at a company whose commercial products have applications related to the personal research interests of that scientist, then everyone can be very happy and free!

What about job security for scientists working in industry or academia?

Doctoral scientists researching at industries or universities proceed down a similar career path, but the challenges and satisfactions are different. **Industrial scientists** must contribute significantly and reliably to their research group; they have job security in the form of a positive administrative judgment about their productive value to research projects at their company. **Faculty scientists** in universities must satisfy the objectives stated in the application for a research grant (i.e., the specific aims), publish important research results that add to their individual reputation, and get their grant(s) renewed. They can have official job security once they become tenured; **academic tenure** is a nominal guarantee of future employment, so freedom of thought is protected. The decision about tenure is "up or out" and is permanent. Job security for scientists working in industry is less embedded in concrete, in that any research project or work assignment can be cancelled for business reasons by administrators; however, good individual researchers are highly valued in R&D labs and do earn some job security. If a project is terminated by administrative decree, valued researchers can be shifted to work on a different project in a different lab.

Although the path for job security has differences at the two locations, I must say here that I know several industrial researchers personally, and all of them seem very satisfied with their research careers, unlike so many of their counterparts in academia.

What are the measures for evaluating research progress in academic and industrial science?

All scientific researchers stand on the shoulders of others who worked earlier but are no longer present. **Research progress in academic science** is recognized as producing concepts, knowledge, and technology that are new and true. **Research progress with industrial R&D** is centered on manufacturing new or improved commercial products so financial profits increase.

New knowledge in basic science often takes years before its significance and value as research progress becomes obvious. On the other hand, developing new technologies and new commercial devices by industrial research and engineering development today often seems to be amazingly rapid.

Essentials to be learned in chapter 4!

1. Research is research, but there are some operational differences between research for scientists in industries or in academia.

2. Basic research is mostly conducted in academia, while applied research is more frequently found in industrial labs as part of their R&D activities with engineers and commercial directors.

3. Money plays a very strong controlling role for research studies at both locations, but via different mechanisms.

4. Research progress in academia centers on producing knowledge that is new and true; research progress in industries aims to increase profits from new or improved commercial products.

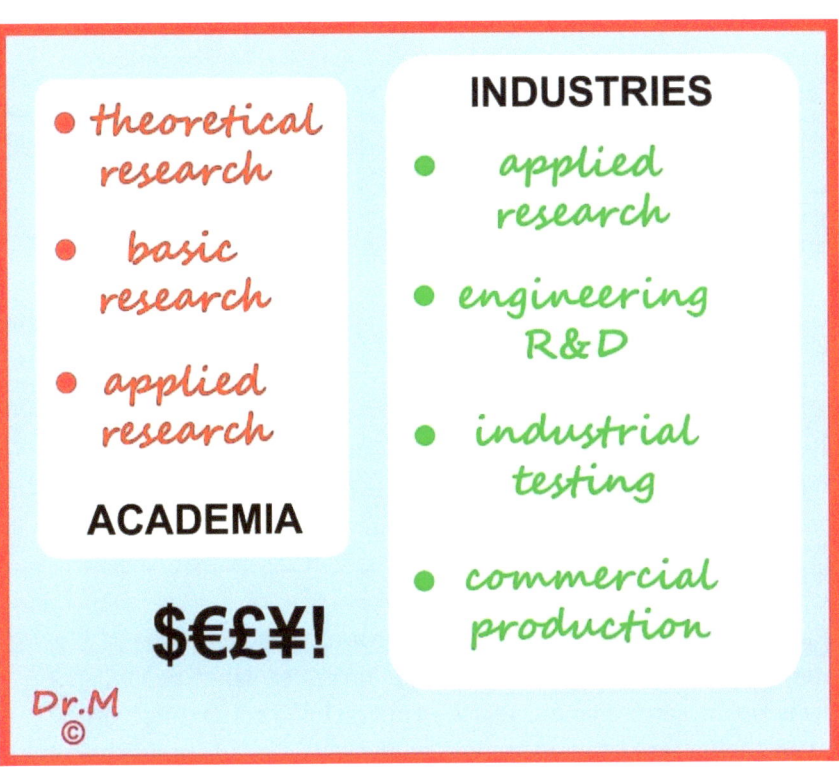

FIGURE 4.2: *Research projects in both academia and industries can be strongly limited by their immense cost!*

CHAPTER 5

Why Is Science So Very Expensive? Who Pays for All the Scientific Research in the United States?

Many scientists working in academic and industrial laboratories will admit that scientific research is very, very expensive. Most scientists in modern universities will also tell you that finding money to enable their investigations is not so easy and that their whole career now depends upon success in getting awarded research grants. This chapter examines why science and research cost so very much and looks at who pays for the expenses of all research projects within the United States.

Why is scientific research so costly?

The total expenses for any research study include paying for the salaries and benefits of all lab employees, purchase of special instruments and research supplies, safe storage and regulated disposal for radioactive substances and toxic chemicals, monitoring of exposure to radioactivity, publication of research reports in professional science journals, submission of applications for new patents, business travel to national and international science meetings and conferences, regulated maintenance and use of research animals, approved utilization of consenting human patients for clinical research studies, synthesis of needed special chemicals and materials, all aspects of specimen preparation, repair of broken instruments, housekeeping of the assigned research lab

SALARIES
  Principal Investigator
  Postdoctoral Fellows
  Research Technicians
  Graduate Students

RESEARCH SUPPLIES
  Routine Lab Supplies
  Special Lab Supplies
  High Purity Water

RESEARCH EQUIPMENT
  Purchase or Rental
  Service Contract
  Repairs and Calibration
  Accessories

TIME
  Conduct of Experiments
  Analysis of Data
  Abstracts for Meetings
  New Manuscripts
  Revision of Manuscripts
  Reports and Forms

MISCELLANEOUS
  Business Travel
  Cell Cultures
  Chemicals Inventory
  Clinical & Hospital Work
  Collaborators
  Computers & Display
  Radioactivity Usage
  Rental Charges
  Research Animal Usage
  Safety Regulations
  Science Meetings
  Storage Regulations
  Toxin Use & Regulations
  Use of National Facilities
  Waste Disposal

COSTS NOT INCLUDED
  Coffee
  Food
  Office Supplies
  Parking
  Telephone Calls
  Transportation

Dr.M ©

FIGURE 5.1: *Why is scientific research always so damn expensive? It's because there are so very many items that necessarily must be paid for within both academic and industrial labs!*

and office spaces, necessary remodeling of the assigned laboratory room(s) so the project's activities can be carried out, and so on. For some scientists, that includes payment for laundry service of lab coats, paper towels used in the lab, parking space, room keys, the salary of the PI, telephone calls, and xerographic copying. This certainly is quite a long list!

What are the salaries for today's professional scientists conducting research?

The wages for doctoral scientists researching **in university science departments** vary for different educational institutions, regions, and countries. During 2016–2017, the average salary for all assistant professors (i.e., not tenured) in the United States was $69,206 per annum;[1] the average salary for all full professors (i.e., with tenure) was $102,402 per year.[1] The corresponding figures for salaries of all doctoral scientists and engineers performing research and development **in industrial labs** are somewhat higher. Star researchers in both academia and industries can receive more than the average figures just stated.

There are two categories of salaries for employment of scientists in academic institutions: 1) **hard-money scientists**, and 2) **soft-money scientists**. The first group have a contractually guaranteed salary (e.g., from a state government), are subject to periodic evaluations possibly leading to obtaining academic tenure, and hold some independence as long as they maintain their external research grant funding.

The soft-money scientists have their salaries paid solely from their research grant(s) and mostly remain untenured. If their grants are not renewed, then their salary and research both stop. Universities like this arrangement very much because they are now businesses where money and profits are everything; with faculty salaries on soft money, they get higher profits from more faculty having multiple concurrent research grants and their necessary costs are notably decreased. If research grant funding stops, that faculty scientist is out on the street, but a new replacement can be hired easily. Soft-money scientists getting multiple research grant awards are pleased to receive higher salaries than do their hard-money counterparts in the same institution.

What are research grants? How are they awarded to university scientists?

For universities, medical schools, and research institutes in the United States, funds to support their research studies come from **external** sources as **research grants** awarded by the government science agencies (e.g., the Agricultural Research Service, Department of Energy, National Aeronautics and Space Agency [NASA], National Institutes of Health [NIH], National Science Foundation [NSF], and so on). Research grants are intended to enable the best proposals for meritorious new or renewed (continued) projects to be conducted. Evaluations for merit are made by **peer review**, where groups of other scientists with expertise and experience in the same field as the proposals being considered are invited to critically evaluate and rate applications. A large professional staff at each agency then meets to make and administer the decisions about funding.

This peer review produces a **relative ranking** or **priority score**. That key number largely determines the ability for any application to be awarded financial support. Unfortunately, despite the huge number of dollars appropriated to pay for scientific research, it was possible in 2016 to fund only around 19–20 percent of all the applications received;[2] thus, many good proposals currently must be turned down completely, and others can be awarded only limited partial funding. Any faculty scientist without a research grant simply cannot do any research. Faculty competition for research grants is very fierce and has become a hypercompetition; it now is a life and death matter for academic scientists!

How are research and development projects funded in industries?

All the large costs for R&D work needed to produce new or improved commercial products in industries are funded **internally**, using business profits of the manufacturing company. Industrial scientists and engineers in principle can be laid off at any time; however, highly skilled workers in industrial labs are greatly valued and receive a nice salary. Industrial scientists don't need the tenure system used in academia because their history of continuing high-quality work reliably contributes to increase the company's profits.

What is the grand total spent for all research in academia or industry in the United States?

The grand total for financial support of all research expenses is truly gigantic! The US federal government provided some 139 billion dollars to support all types of academic and other research activities with research grants and other mechanisms in 2015.[3] Industrial companies spend even more billions for their research and engineering development activities; their $341 billion in 2014[4] exceeds the funding awarded by the national government for academic research projects.

Who actually provides the billions of dollars paying for research in the United States?

All money awarded by the US government as research grants comes from taxes directly paid by the US taxpayers! Almost all money used to support industrial research and development by commercial businesses comes from their pool of profits; those internal funds are generated by sales to individual customers and other companies, meaning that members of the public also pay for all the industrial R&D expenses! Thus, *all people in the public, whether rich or poor, young or old, male or female, pay for scientific research and industrial developments!* And that includes you and me!

Those realizations have an important derivative that is not realized by most people: since the public pays for all scientific research, then they are the chief owners of science! That leads to two derived conclusions. First, it is good for everyone in the public to know at least a little about science, be able to see what research is supposed to do, and accept that individual research scientists are just fellow people who have lots of special education and training. Second, scientists must participate more readily in educating the public about their work; ideally, the public should then become more interested in scientific research so that there would be a more active ongoing dialogue between people and scientists.

Essentials to be learned in chapter 5!

1. Scientific investigations are so costly primarily because there is a very long list of necessary expenses for any research project.

2. Billions of dollars are spent each year for research by faculty scientists via research grants from governmental science agencies.

3. An even greater sum is spent each year by industries for research and development by their scientists and engineers.

4. All research is directly or indirectly supported by the public!

5. Scientists and people in the public should more actively communicate with one another!

References

1. "Visualizing change: The annual report on the economic status of the profession 2016–2017," *American Association of University Professors*, 2017, https://www.aaup.org/report/visualizing-change-annual-report-economic-status-profession-2016-17.

2. "NSF Science & Engineering Indicators, 2016. Table 5-22, NIH and NSF Research Grant Applications and Funding Success Rates: 2001–14, *National Science Foundation*, 2016, http://www.nsf.gov/statistics/2016/nsb20161/uploads/1/8/tt05-22.pdf.

3. "Table 1, Federal Budget Authority for R&D and R&D Plant, by Budget Function, Ordered by FY2015 R&D and R&D Plant Total: FYs 2015–17," *National Science Foundation*, 2017, http://www.nsf.gov/statistics/2017/nsf17305/pdf/tab1.pdf.

4. R. M. Wolfe, "Businesses Spent $341 billion on R&D Performed in the United States in 2014," *National Center for Science and Engineering Statistics,* 2016, http://www.nsf.gov/statistics/2016/nsf16315/nsf16315.pdf.

FIGURE 5.2: *Although scientific research and engineering development in the United States cost many billions of dollars annually, that huge sum is never really enough! However, the simple truth is that all the numerous scientists really do not need multiple research grants!*

CHAPTER 6

What Does Grouping Do for Research Projects? What Is Little Science Versus Big Science?

When all is said and done, science is people! The standard image of a "very famous scientist" shows a bald old man with a long white beard working all by himself or an old lady with lots of white hair working all by herself! Such caricatures differ totally from reality, where each real famous scientist has associated research workers; indeed, laboratories in academia usually house grad students, postdocs, and technicians, all working under the direction of one doctoral scientist (see chapters 2 and 3). These coworkers conduct experiments in one or several rooms, constituting one **research laboratory**. That fundamental unit for research can be expanded to form groups of different sizes, complexity, and cost. This chapter examines the different sizes of research groups within academic and industrial institutions, and explains the pros and cons of groupings for research operations comprising Little Science or Big Science.

Why are small research groups so often formed in academia?

Today, basically all new faculty scientists in academia start their research careers at the individual level, but soon advance to form a **small research group**, especially after their first research grant is received. The advantages of this grouping are several. First, the boss of the lab only has two hands and cannot be at two different places at the same time; by adding grad students and technicians to the lab staff, that problem can be neutralized. Second, if several faculty scientists can pool their efforts and collaborate to do some research together, the amount of funding increases since each such faculty member can obtain his or her own research grant; in addition, this small research group can try to obtain a new grant supporting the combined collaborative research effort.

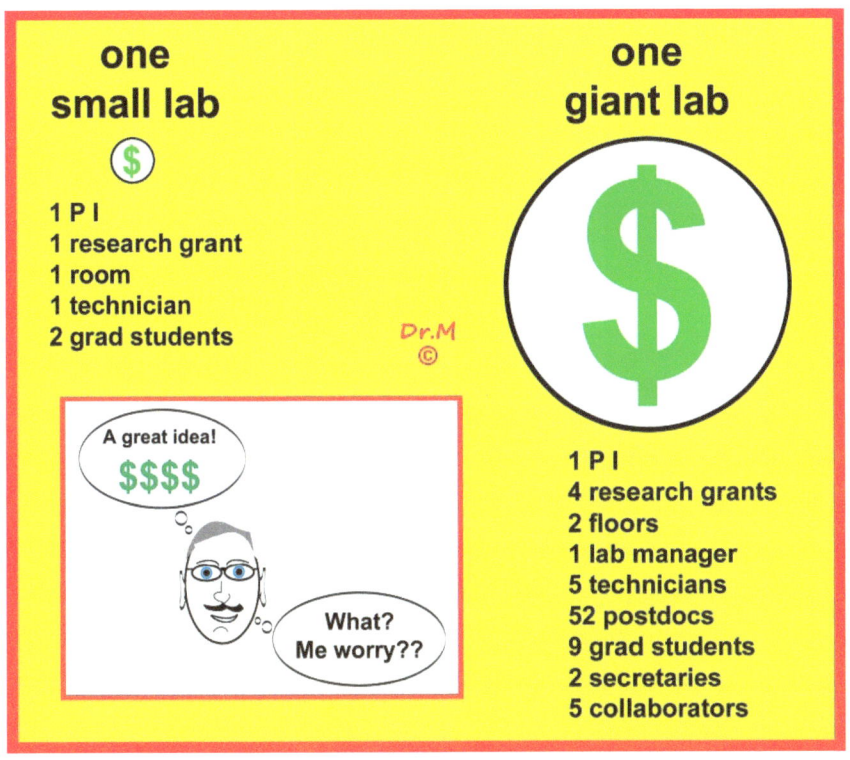

FIGURE 6.1: *The number of staff workers and the necessary costs for conducting research studies vary directly with sizing of the laboratories! Indeed, each giant lab can realistically be compared to an entire fleet of numerous smaller labs in universities!*

Third, lab operations become more efficient, and work assignments are more flexible (i.e., if there are three or more technicians in a small group, each can either work separately with that one scientist providing their salary, or all can work together with one scientist in the group whose current experiment requires many hands). Fourth, any small group with several doctoral scientists has more brains, meaning that there will be more frequent interactions, more internal criticisms, more new ideas, more questions, and a wider range of expertise.

How do larger research groups form and develop in academia?

When a small research group becomes very productive, it naturally will tend to expand into a **large research group**. Everything is scaled up, but continues to be directed by one group leader, the principal investigator. Larger groups can be more flexible and more reliable for getting a research project fully completed on time. Any large research group can complete research projects in a fraction of the time it takes for one small research group to finish everything; having more lab staff means that new experimental data can be acquired more readily, and projects will be completed much sooner. Preliminary data also will be gathered more quickly by a large group, thereby allowing it to apply for a new research grant before competitors can. It is almost suicidal for any small research group to compete directly against a large research group.

In practice, some large and successful **giant research groups** in academia can grow so big that they fill several floors or even an entire building! The biggest groups can include one hundred postdocs! Those then often are termed **research factories** by other scientists, because with their very large budgets, numerous staff workers, huge size, and enormous productivity, they can simply overwhelm any research competition.

How are groupings used for industrial research and development projects?

In general, research studies in industries often use large research groups for applied research and engineering developments. Within very big companies, investigations can use multiple research groups, with each working in coordination on different related aspects of one project. Giant corporations can organize a number of separate large research groups working in coordination at widely separated locations; this can be termed a **dispersed large research group**. The industrial groupings effectively combine the advantageous features of both smaller and larger operations.

Is growth of academic research groups always good?

The size of research groups in academic institutions can have either positive or negative effects. I have personally seen several very big research centers that were admirable in all respects; they were hugely productive, highly efficient, and had the latest special research instruments on hand and ready to use. However, not everything necessarily improves with a larger size of research operations! Some have questioned whether the same number of dollars awarded to one giant research group would not produce a better quantity and quality of results if they were divided and then awarded to a number of small research groups. Nobody knows the answer to this provocative question!

Research advances are ultimately rooted within the output from **individual scientists** who supply creative ideas, new concepts, personal insight, research breakthroughs and new directions, special talents, unconventional approaches, and so on. **Individualism** can tend to become submerged within research factories or giant research groups. In such cases, doctoral scientists working within the biggest groups are actually highly paid specialized research technicians; their input of creative new ideas and unconventional interpretations of data decreases, meaning that the reasons why they spent so many years working to get a PhD vanish.

What is Little Science versus Big Science?

Scientists are employed for projects conducted at the level of either Little Science or Big Science. Both categories are useful, but each is better suited for certain types of projects. **Little Science** includes all the common research projects within academia; these are supported by all the standard research grants (see chapter 5). Typically, they take a shorter time to reach completion, require a smaller amount of money, and can excel with breakthrough discoveries, creative new concepts, and novel findings.

Good examples of projects in **Big Science** include the planning, construction, and operation of a new multiuser synchrotron research facility, a new space telescope, and a new large or very advanced terrestrial telescope. Other Big Science projects include giant research surveys involving humans or specimens from multiple continents. Big Science projects often take many years or even decades to reach completion. All projects in Big Science are unique, need giant funding, and require involving very many doctoral scientists, numerous engineers and myriad research staff. Big Science projects are funded as a giant group activity with an extensive administrative component; they are so costly that some can be funded only by a multinational consortium. In some cases, Big Science projects can evolve to become formalized for ongoing research as a national or international facility (e.g., NASA, the National Aeronautics and Space Administration [https://www.nasa.gov]).

The enormous cost of Big Science projects is justified by the expected significance of the special data produced, the large number of scientists involved, and the number of people affected by the research results. For NASA, its space explorations, production of advanced telescopes on Earth and in space, other special instrumentation developments, technological innovations, and many useful by-products strongly justify its huge cost.

The distinctions in the work done by Little versus Big Science are analogous to construction of a residential house versus building a skyscraper. The prominent differences in time needed, amount of money spent, and number of skilled construction workers employed correspond nicely to features for the two categories of projects in science.

Teamwork is vitally important for modern scientific research!

Working together as a **research team** is extremely valuable for research projects within both Little and Big Science. **Teamwork** will increase both the quantity and the quality of new findings. The importance of the team approach has been much more recognized in industries than in academic institutions. However, the **"industrial model"** for scientific research studies with a team approach has recently been utilized in a few new centers for lab-based biomedical research funded by philanthropists.[1,2]

Essentials to be learned in chapter 6!

1. Only rarely do any modern doctoral scientists conduct research all on their own; lab staff are needed to make research progress!

2. Research groups of small size can expand into larger units, thereby scaling up the several benefits of grouping.

3. Research projects in Little Science include all investigations conducted by faculty scientists holding a standard research grant.

4. Big Science includes special projects with numerous workers, many years of work, and a giant budget; they are often unique.

5. Teamwork by all scientists and research coworkers within any size of lab has vital importance for producing solid results.

References

1. "Dr.M, 2015, "A Jackpot for Scientific Research Is Created by James E. and Virginia Stowers! Part II. The Stowers Institute Is a Terrific New Model for Funding Scientific Research!" Dr.M on Science, Research, & Scientists,

2015, http://dr-monsrs.net/2015/10/20/a-jackpot-for-scientific-research-is-created-by-james-e-and-virginia-stowers-part-ii-the-stowers-institute-is-a-terrific-new-model-for-funding-scientific-research/.

2. Dr.M, 2016, "A Dramatic Individualist, Paul G. Allen, Is a Major Benefactor of Scientific Research!" Dr.M on Science, Research, & Scientists, 2016, http://dr-monsrs.net/2016/04/06/a-dramatic-individualist-paul-g-allen-is-a-major-benefactor-of-scientific-research/.

FIGURE 6.2: Differences in characteristics between Little Science and Big Science are very dramatic!

FIGURE 6.3: *Dr.M says: "Hoorah! We're at the halfway mark! There are many interesting items ahead for you to view, read, examine, and laugh at!"*

CHAPTER 7

On the Nobel Prize and Other Big Honors in Science!

All of us seek some recognition and rewards for competing with other people or nature, no matter our age, ambition, gender, job, personal interests, or skills. Some seek money, others want fame or power, and yet others aim to get personal pleasure. Scientists working in a university or an industrial center desire deep down to achieve widespread recognition from other researchers; they want to be seen by other scientists as being a leader of research and "a **great scientist**." The supreme honor for scientists is to win the Nobel Prize or one of the newer very big awards.

This chapter explains what the Nobel Prize is, describes several more recently established multimillion dollar prizes for important research discoveries, and discusses the obvious question of what good are the multitude of other scientists who are *not* winning any awards?

The foremost prize for excellence in scientific research was originated by Alfred Nobel, a researcher and industrial magnate who invented dynamite!

Alfred Nobel became very wealthy from his invention of dynamite and other explosives. Born in Sweden in 1833, he received only limited education in schools. He was a prolific inventor and had over 350 patents by his death in 1896. Much of his very large industrial fortune was bequeathed to establish the Nobel Prizes in science; he specifically stated that they are to be awarded to "those who during the preceding year have conferred the greatest benefit to mankind." For further biographical information about Alfred Nobel, see:
http://www.nobelprize.org/alfred_nobel/.

FIGURE 7.1: Portrait of Alfred Nobel in late 1800s by Gösta Florman. Common Domain Wikimedia image (https://commons.wikimedia.org/wiki/File:Alfred Nobel.jpg).

Nobel Prizes are awarded to individuals (Nobel Laureates) in each of the three major branches of science (physiology or medicine, chemistry, and physics). Additional Nobel Prizes have more recently been established in economic sciences, literature, and peace. The first Nobel Prize in science was bestowed in 1901, and a very long chain of annual awards for outstanding research continues to today. In the 116 years that have passed, a total of 911 Nobel Prize Laureates (in all topical categories) have thus far been honored. It is amazing to note that a very few scientists, such as Linus Pauling, have even been awarded two Nobel Prizes! The youngest scientist to win a Nobel Prize was Lawrence Bragg, who was only twenty-five years old! The winners come from many different countries and include forty-nine females (http://www.nobelprize.org).

Selection of the honorees is conducted and administered by the Royal Swedish Academy of Sciences, the Nobel Assembly of the Karolinska Institute, and the Nobel Foundation. All of these are based in Sweden. The newest Nobel Prizes are bestowed annually by the royal ruler of Sweden at special ceremonies during "Nobel Week" in December. Each new Laureate presents a Nobel Lecture, receives the Nobel Medal, and is given a special document stating their substantial monetary reward. Most of the Nobel Laureates in science declare that receiving this supreme honor is truly a once-in-a-lifetime experience.

The Nobel Prize website is a very valuable repository for historical information and materials about scientists!

The fascinating website for the Nobel Prize (see http://www.nobelprize.org/) presents much useful information about all the past winners and the newest Laureates, explains how their research accomplishments are especially important, and offers videos with interviews and details about their lives and research work. The Nobel Lectures from Nobel Prize Week are recorded and made available there. Videos with educational features are also prominently displayed on the Nobel website.

I find it fascinating to watch videos showing some extremely famous but very ancient pioneering scientists who won a Nobel. If you want to try this glimpse into science history yourself, simply find the name of a world-famous Laureate in whom you are interested and then see if any videos are available for that person on the Nobel Prize website.

Who wins, and who does not win, a Nobel Prize? Why?

Looking over all the Nobel awardees, these highly selected individuals feature diverse personalities, dedication to researching, and a flair for experimental investigations. They clearly have strongly influenced both science and society. No one doubts that all the Nobel Laureates deserve their honorary recognition, but it remains quite unclear why certain other renowned investigators have *not* also been honored by a Nobel.[1,2] Some instances are due to the official rules (e.g., a Nobel Prize cannot be bestowed upon anyone posthumously or to more than three individuals working on the same project); in other cases, no logical reason explains this longstanding question.[1,2] Certain disciplines of science have not yet been accorded any Nobel Prize winners. These absences are part of the reasons for establishing new honors for scientific researchers at the level of the Nobel Prizes (see below).

Several additional very big prizes now are offered for excellence in research by scientists!

The **Kavli Prize** was established in 2008 by the late Fred Kavli (1927–2013), a very successful physicist, entrepreneur, industrialist, and philanthropist. The Kavli awards (see http://www.kavliprize.org) honor research excellence only in astrophysics, nanoscience, and neuroscience, whereas the Nobel Prizes can be awarded to outstanding researchers in any different part of science. The Kavli Prizes are awarded every two years and are now widely regarded as having the same very high level of prestige and distinction as do the Nobel Prizes. The Kavli Prizes are bestowed to the honorees by the royal ruler of Norway at special ceremonies during Kavli Prize Week in Oslo, Norway.

Several other prestigious big honors for excellence in research have also been established recently. The **Breakthrough Prizes** in fundamental physics, life sciences, or mathematics reward each honoree with several million dollars (https://breakthroughprize.org/); these financial rewards are notably larger than the amount given for a Nobel Prize.

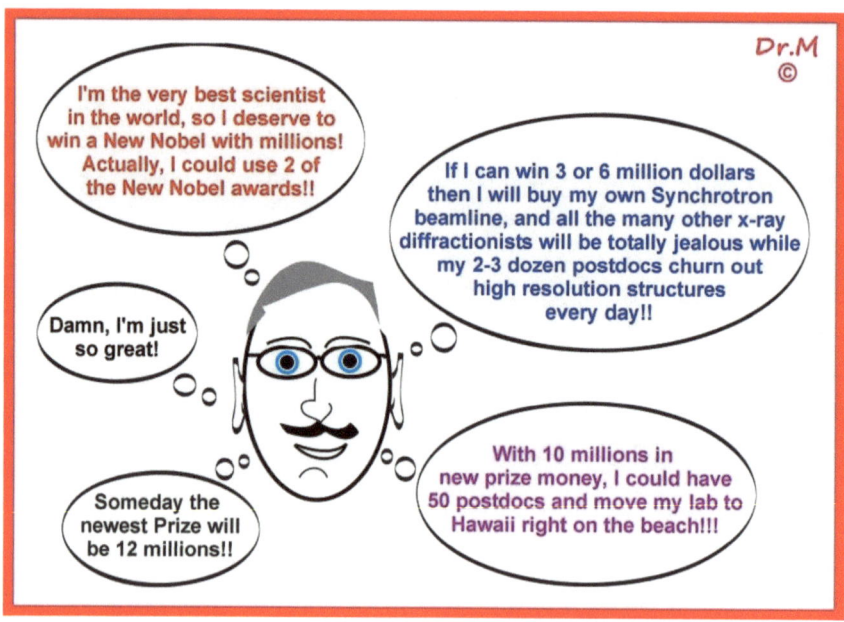

FIGURE 7.2: *Very many scientists dream about winning a giant prize with a huge pile of dollars, but only a few ever see their wish fulfilled!*

The several Breakthrough Prizes are awarded at a star-studded, widely publicized, very large extravaganza intended to promote public awareness and appreciation for scientific research.

Other big new megaprizes include the Tang Prize in Biopharmaceutical Science and the Queen Elizabeth Prize for Engineering. Further information and videos about these new grand honors for scientists are very nicely reported by Zeeya Merali in "The New Nobels" (https://www.nature.com/polopoly_

fs/1.13168!/menu/main/topColumns/topLeftColumn/pdf/498152a.pdf).

What good are all the scientists not winning any big prizes?

Of the many thousands of doctoral scientists now researching in academia and industry, only a very few will ever win a big prize for excellence in research. All the others still must be recognized as being valuable! The multitude of **nonwinners** produce the great bulk of research reports published in science journals each year. Many are highly dedicated teachers of science to students and the public. Some also are active with administration, advising legislators and government officials, consulting with corporations, serving on national committees for science societies, working with media projects to bring science to the public, and so on.

In addition, an increasing number of young doctoral scientists are now found in nonscience jobs dealing with administrating, consulting, designing advertisements, developing new technologies, establishing new commercial products, evaluating science stocks for mutual funds, inventing, running a new small business, participating in science service operations, and so on; those scientists will not win either a Nobel Prize or any of the other great honors, but they truly are very valuable for science, society, and technology!

Essentials to be learned in chapter 7!

1. The Nobel Prize was established 116 years ago and continues to be the very highest honor for excellence in scientific research.

2. Several wonderful new megaprizes for excellence in research recently have been established, which reinforce the worthy concept that society should give more appreciation to outstanding research studies by doctoral scientists.

3. The multitude of professional scientists never winning any big honors still benefit science, society, and technology; they publish the great bulk of new knowledge coming from scientific research and contribute much to science education and local or national service.

References

1. Z. B. Lord, "10 Scientists Robbed of a Nobel Prize," *Listverse*, 2011, http://listverse.com/2011/11/05/10-scientists-robbed-of-a-nobel-prize/.

2. L. S. Hwang, "5 Great Scientists Who Never Won a Nobel Prize," *World Science Festival*, 2014, http://www.worldsciencefestival.com/2014/10/5-great-scientists-never-won-nobel-prize/.

FIGURE 7.3: *Gigantic prize rewards certainly are much appreciated by the winning scientists, but can also have a few problematic side effects (e.g., some celebrated winners can become less active)!*

CHAPTER 8

Let's Meet Some Real Scientists and Inventors!

Probably no readers of this book have ever spoken to a real living scientist, even though one could be living next door! What kind of people are scientists? Those researching in academia or industry might be quiet or loud, friendly or dour, young or old, wear a suit and tie or a T-shirt and jeans! Female scientists do have husbands and children, sometimes wear dresses, and can chair a department! Many scientists also are inventors, but most inventors are not scientists.

Most scientists have some interesting and amazing tales to tell about their activities with research in a lab or with field studies in hospitals, nature, oceans, or space. To let you get to know some of these people better, this chapter presents brief accounts about a few individual inventors and scientists; it also suggests how you can try to meet a research scientist in the flesh without worrying that they might bite your head off! The inventors and scientists selected here are certainly not the only ones that are interesting. Try to meet one yourself, and you will see what I mean!

We first will meet a very productive modern inventor!

The late **Artur Fischer** (1919–2016) invented and developed several quite different and very useful devices, one of which is extremely well known to billions of people. He only had a limited education in schools; this characteristic is surprisingly shared with some other highly famous inventors! Fischer developed his creative ideas by working with his hands and eyes in his workshop; he was very big on tinkering to make changes to advance his innovations

FIGURE 8.1: Scientists are just like other people! Typically, they are very interesting to meet because they have had many adventures while conducting research investigations! Do try to get to talk to one all by yourself!

(i.e., he did experimental development by trial and error). A business to commercialize his inventions was started in 1979 and has continuously grown and expanded; today, it is headed by his son, Klaus, an award-winning engineer. The Fischer Group of companies now has many factories and branches located in several different countries (for an overview of their activities, see https://www.fischer.group/en/about-fischer).

> *"When I paint a picture I start out with a blank canvas, and it's just like when I'm in my workshop!"*
>
> *"It's okay to make mistakes!"*
>
> *"I look at my life as a gift, because there is nothing more wonderful than inventing!"*
>
> Quoted from **ARTUR FISCHER** in the 2014 video issued by the European Patent Office at: https://www.youtube.com/watch?v=hmkZ_ipbY90

The greatest invention by Artur Fischer is the **screw anchor**, which permits objects such as mirrors, paintings, or shelves to be hung onto walls and solid surfaces. Screw anchors are available today in myriad colors, sizes, and types, and are used universally by carpenters and homeowners around the world. Specialized screw anchors are now even used by orthopedic surgeons to repair fractures producing fragmented bones; one really good invention often evolves into some others! It is estimated that millions of these very simple devices are sold every week.

Fischer also invented educational toys! His second-most well-known invention is mainly used by children (and also by their parents!), and can be purchased as sets of different small components to be used for assembly and construction of complex model buildings (see http://www.fischertechnik.de/en/Home/products.aspx). This commercialized invention enables youngsters to incorporate various motors and special features into their own designs. The different component pieces relate to one another in all three dimensions and enable children to add technology into their construction of miniature buildings, factories, or machines. This toy is clearly both fun and educational for all youngsters!

Note that this outstanding inventor did not hold a PhD and was not a scientist, but his creative visions progressed into science education. His limited education probably resulted in his native curiosity and creativity *not* being squelched in the usual antieducational classes we all endured in primary and secondary schools. To see this inventor in action, take a look at two good videos: "Artur Fischer, Wall Anchors, Synchronized Flash, and Many More" (https://www.youtube.com/watch?v=Ke5K4S-Cjj8), and "Artur Fischer in His Own Words—Winner of the European Inventor Award 2014" (https://www.youtube.com/watch?v=hmkZ_ipbY90).

Now here's a physicist, inventor, educator, and industrialist!

Edwin H. Land (1901–1991) is renowned as the inventor of instant photography, which he developed commercially as Polaroid cameras and black-and-white or colored Polaroid films. This enabled rapid production of both negatives and hard-copy photographic prints before the later invention of digital cameras. Land worked at his own research while he was still a beginning college student; enrollment in classes stopped before he finished his degree, but Land's efforts with creative research never ended even after he achieved enormous success with business activities!

Land worked in laboratories set up inside his Polaroid Corporation, which at one time had over 10,000 employees. He actively supervised teams of doctoral scientists and engineers for developing new commercial products, such as the colored version of Polaroid instant imaging. Land also used his great creativity to develop an alternative theory explaining the human vision of colors. This inventive and driven man was always looking into the future and held over 3,000 patents. His enthusiasm for research notably benefited his industrial operations, his country in wartime, and many ordinary people.

> *"It is a curious property of research activity that after the problem has been solved the solution seems obvious."*
>
> **Quoted from EDWIN H. LAND in A-Z Quotes (http://www.azquotes.com/author/8443-Edwin_Land)**
>
> *"Land was legendary for his eccentric and exhausting work habits. He would lock himself in the lab for days on end. Land's assistants would be scheduled in shifts … ."*
>
> **Quoted from TONY LONG (1947) in Wikiquotes (https://en.wikiquote.org/wiki/Edwin_H._Land)**

One of Land's side interests was science education. At Harvard College, he designed and activated a special program for all undergraduate students where they worked for a full year on research within a faculty lab in addition to taking classes. To see Edwin Land in action, take a look at two very good videos available on the Internet: "The Polaroid Genius Who Re-Imagined the Way We Take Photos" by BBC News Magazine (http://www.bbc.co.uk/news/magazine-21115581), and "The Long Walk (1970)" by the Polaroid Corporation (https://www.youtube.com/watch?v=zbmq9R0dtVg).

Next, we'll meet a very creative and unconventional scientist who is a celebrated and individualistic chemist!

Kary B. Mullis is an experimental chemist who is very widely renowned for his discovery of the polymerase chain reaction, which made it readily possible to sequence genes. He won a Nobel Prize in Chemistry in 1993 for that key breakthrough discovery. The story of how that advance was made and led to the development of molecular genetics, a new subbranch of bioscience, is described in a very personal way; many amazing incidents occurring during Mullis's adventures in life and in research, are described in his autobiographical essays, "Biography" and "Making Rockets" (https://www.karymullis.com/biography.shtml). Anyone who dislikes science and research will still be fascinated to read this true story!

Kary Mullis has a unique and very distinctive personality and he pays no attention to political correctness! He is enthusiastic about both creative thoughts and creative research. I urge all to watch the video "Interview with Kary B. Mullis" at http://www.nobelprize.org/mediaplayer/index.php?=428. His great sense of humor is wonderfully apparent in a TED video presentation, "Sons of Sputnik: Kary Mullis TEDxOrange Coast" (https://www.youtube.com/watch?v=iSVy1b-RyVM). Dr. Mullis is certainly an outstanding example of how some very individualistic scientists can succeed at research

> "At Dreher High School we were allowed free unsupervised access to the chemistry lab. Today we would be thought of as a menace to society."
>
> "Cetus hired me ... I worked long hours and enjoyed it immensely."
>
> "... my first PCR paper was rejected" (by *Science* !)
>
> Quoted from KARY B. MULLIS: 1993 Nobel Lecture (http://www.nobelprize.org/nobel_prizes/chemistry/laureates/1993/mullis-lecture.html)

and yet still remain very strong individuals and still achieve great recognition (see "Curiosity, Creativity, Inventiveness, and Individualism in Science," http://dr-monsrs.net/2014/02/25/curiosity-creativity-inventiveness-and-individualism-in-science/)!

Here's a pioneering cell biologist and experimental pathologist!

Marilyn G. Farquhar is widely renowned for making several key research discoveries in modern cell biology. With pioneering research on the subcellular pathology of the kidney during different disease states, she established the solid basis for using electron microscopy of kidney biopsies to make clinical diagnoses of renal diseases. Professor Farquhar's many research studies in cell biology have resulted in detailed knowledge about the structural and functional activities of several subcellular organelles and about the dynamics and regulation of certain proteins within the cytoplasm.

Dr. Farquhar actively continues to be a research leader and has received many honorary awards and medals for her investigations. She also is revered for her erudite teaching lectures. She has given back to science by serving as the elected president of the American Society for Cell Biology, and as the chair of her Department of Cell and Molecular Medicine at the University of California (San Diego). Although there appear to be no videos on the Internet featuring Dr. Farquhar, some impression of her dynamic personality and ongoing enthusiasm for researching can be gained by reading a recent interview report by C. Sedwick, "Marilyn Farquhar from the Beginning" (http://jcb.rupress.org/content/203/4/554.full.pdf).

> " ... for my PhD thesis project I studied secretiom in pituitary cells. It was an exciting time ... because whatever you looked at was new".
>
> "... my imprinting and commitment to kidney research came as a post-doc ... ".
>
> MARILYN G. FARQUHAR in *Kidney International* 64 (2003) 1943-44.
>
> "It was a lomng trip from the identification of megalin as a target antigen of an autoimmune disease to understanding the functional significance of its trafficking route".
>
> MARILYN G. FARQUHAR from interview by C. Sedwick in *Journal of Cell Biology* 203 (2013) 554-555.

How might you be able to meet a real living scientist in person?

Probably nothing will teach you more about science, research, and scientists than meeting and talking to a real professional researcher in person! Most doctoral scientists like to tell other people what they are investigating and relate stories about their activities in conducting research experiments. Finding a scientist to talk to is not so easy, but if you ever have the chance to do that, please go ahead! By visiting the annual meeting held by any national science society, you will have numerous opportunities to hear, see, and talk to research scientists of all ages; although you might be required to register and pay for visiting that gathering, often you can simply walk in and look around without paying.

To find a scientist, you could look to see if one is giving a public presentation nearby (e.g., at a university department). You also might ask local science teachers if they know any professional scientists not too far away who you could meet, or know of a scheduled seminar or public presentation at a nearby location. If so, ask that scientist if you can meet him or her on some later day for an hour or so; ideally, you might then also be given a tour of the scientist's research lab. Be sure to read several of that scientist's recent research publications before any meeting, so that you will be better able to understand the conversation about his or her research work. After you have met and talked to a real living scientist, I have no doubt that you then will be astounded to realize that *brainy scientists are distinctive individuals, but really are only other people*! They have some of the same challenges, experiences, and problems with everyday life as you encounter, plus some others!

Essentials to be learned from chapter 8!

1. Scientific research can be an adventure, with creation, discovery, and exploration!

2. Most professional scientists as individuals are very dedicated to their research investigations, have a large and active sense of curiosity, feature a creative mind, and always come up with many questions regarding science, research, and the universe.

3. When all is said and done, scientists are only other people! Individual scientists are not clowns, they are not mad, they work hard, they do make some mistakes, and they are valuable to both science and society!

FIGURE 8.2: *Most people, even modern science teachers in primary and secondary schools, have never in their lives met and talked with a real living research scientist!*

CHAPTER 9

Could Any Modern Research Scientists Actually Be Criminals?

Although science continues marching forward, it does have several very difficult problems. One of the biggest is research fraud and dishonesty by professional scientists. Almost all research scientists seem to be honest, but cheating, corruption, and lying within both academia and industry are increasing. Most people in the public continue to believe that all scientists are completely honest simply because they are ignorant about this very serious issue in today's scientific research.

Most readers might guess that the answer to the question posed in the title must be "no," but the true answer, regrettably, is a big "yes"! This chapter briefly examines why **research misconduct** and poor professional ethics are found in modern science.

A brief introduction to research misconduct!

Cheating at research includes such unethical behavior as changing experimental research results in order to support some desired conclusion (e.g., a new chemical medication can be wrongly labelled as safe for human usage because research results showing bad side effects are not reported or published). Some other examples of research dishonesty include failing to cite relevant results or conclusions published earlier by other scientists, making up completely imaginary data and publishing them, outright theft of results or materials from other scientists, plagiarism for new publications, removing some outlying data points so an averaged figure is made much better, and so on. Although some cheating has been going on in science for a very long time (centuries!), any dishonesty by modern scientists runs against the most fundamental goal of all scientific research: namely, to find out exactly what is true.

FIGURE 9.1: *Attempts to make scientists be honest run headlong into the amazing fact that very many in the public and businesses now believe that some dishonesty is OK, especially if it is not detected!*

A typically sad example of unethical behavior received international attention in 2015–2016, concerning **research misconduct by Dr. Haruko Obokata**. Lab research results published by this young scientist appeared to show the discovery of a new and spectacularly easy way to generate stem-cells, the early precursors to the fully specialized mature cells in tissues. Other stem-cell researchers were amazed, but could not duplicate her published results; questions then arose about the honesty of her data. Later, an official investigation found that she could not duplicate her own published results in several carefully monitored attempts. The spectacular research discovery reported by Dr. Obokata thus must be viewed as a falsity. To date, she has admitted little. Punishments for the research dishonesty by this young scientist were discharge by her employer, obtaining a bad reputation with other scientists, and personal shame (one research official involved with her case committed suicide!).

Dr. Obokata's sad story makes very interesting reading for both ordinary people and working scientists. "What Pushes Scientists to Lie? The Disturbing but Familiar Story of Haruko Obokata" by John Rasko and Carl Power is an excellent place to start learning more about criminality in science and research (see https://www.theguardian.com/science/2015/feb/18/haruko-obokata-stap-cells-controversy-scientists-lie/).

How much cheating occurs in today's scientific research?

Although most working scientists will agree that some cheating at research does happen, nobody can give an actual number for its occurrence! Since it is always very hard to prove allegations of dishonesty in scientific research, many instances are probably never counted. The actual number might be surprisingly high! **Dr. Marcia Angell**, former editor in chief of the prime medical science journal, *The New England Journal of Medicine*, stated in 2009 that "It is simply no longer possible to believe much of the clinical research that is published..." (see "Drug Companies and Doctors: A Story of Corruption" at https://www.nybooks.com/articles/2009/01/15/drug-companies-doctorsa-story-of-corruption/). **Dr. Richard Horton**, chief editor of the very prestigious UK medical journal, *The Lancet*, shockingly stated in 2015 that "...much of the scientific literature, perhaps half, may simply be untrue" (see "Offline: What is Medicine's 5 Sigma" at http://www.thelancet.com/journals/lancet/article/PIIS0140-6736(15)606961/fulltext). Additionally, the view that cheating at research now is increasing is supported by all the media reports of new scandals and lawsuits alleging criminal misconduct by scientists in both academia and industries.[1]

What are the effects of dishonesty and criminality in science?

Most people in the public have a very high esteem for professional scientists and would never dream that some are actually liars and cheaters. They are absolutely amazed whenever it is proven that some research scientist is dishonest and unprofessional. Any dishonesty in research studies lowers the integrity of science, hinders the outcome of future research studies based on published falsified results, and decreases the public regard for scientists. Today's scientific researchers typically start a new project by assuming that the results published in previous investigations are valid; this very general assumption is now increasingly risky. Upon finding that research results reported in publications are not reproducible, those eager scientists already have wasted much time and money; detailed plans for their new experiments must then be changed or abandoned. Intentional deceit or misrepresentation in science are nothing less than a criminal act!

Why would any professional scientist ever be dishonest?

Scientific research is often difficult, takes up much time, and always proceeds with no guarantee that experiments will give the expected or desired results. The chief cause for **dishonesty by university scientists** is that their enormous job pressures can overwhelm them (e.g., getting and renewing research grants, producing more research publications more quickly, making more breakthrough research findings, and publishing more articles in the most prestigious journals). All of these pressures arise from the conversion of science at modern universities into a business venture where money is everything; research at universities is now viewed by universities mainly as a wonderful means to increase their financial profits via acquiring more research grants. For some faculty scientists, getting the money from grants becomes more important than finding new truths; they try to take the easy way out, but one small wrong that remains undetected can easily lead to larger ones!

The chief cause of **dishonesty by industrial scientists** is the pressure from a corrupt employer to only produce research results making their commercial products look better. Some **industrial companies**, but not all, avidly seek to add to their financial profits regardless of how many people might be hurt by their falsified research investigations. If an honest researcher obtains data showing that a new pharmaceutical agent being developed has a bad side effect, the manufacturing company might openly request that the employee leaves those negative research findings unannounced and unpublished. Such corrupt manufacturers clearly value their profits above all else and do not care what happens to innocent patients who take their falsely approved harmful medical products. For the honest research scientist, this results in having to decide whether to comply or not.

To go ahead and publish anyway takes extraordinary personal courage and dedication to the true aim of science, since it means ending your employment, having difficulty conducting your future research studies, getting a bad reputation, and having financial difficulties for the rest of your life. A breathtaking exposé about dishonesty by medical researchers and pharmaceutical companies is nicely given by **Dr. Peter Wilmshurst**, a truly heroic medical scientist who is recognized today as being a very courageous and persistent **whistle-blower** (see "Whistleblowers in Science Are Necessary to Keep Research and Science-Based Industries Honest!" at http://dr-monsrs.net/2015/10/27/whistleblowers-in-science-are-necessary-to-keep-research-and-science-based-industries-honest/).

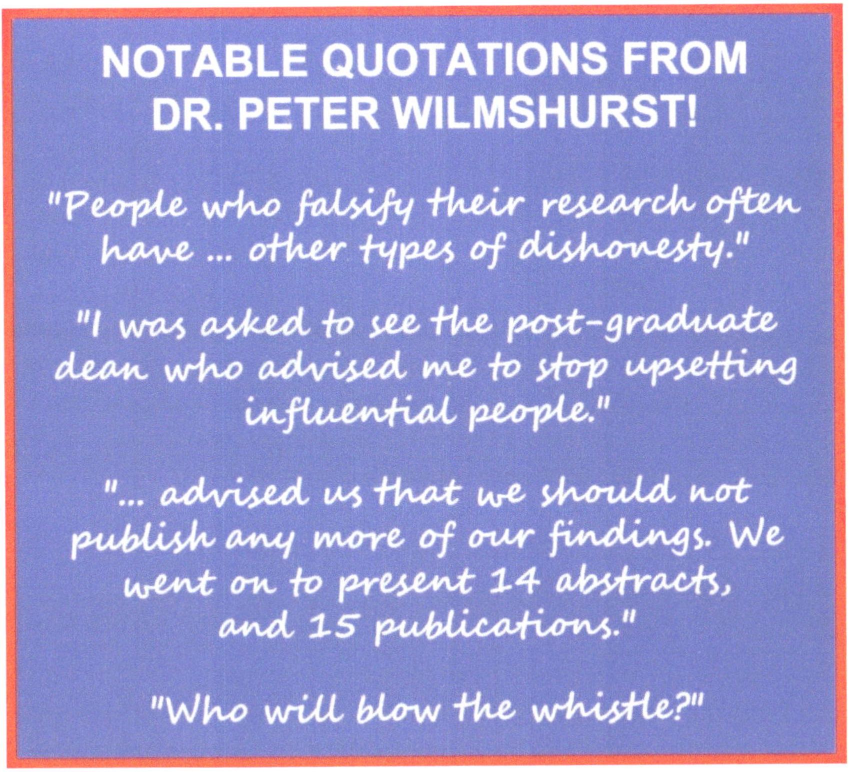

> **NOTABLE QUOTATIONS FROM DR. PETER WILMSHURST!**
>
> "People who falsify their research often have ... other types of dishonesty."
>
> "I was asked to see the post-graduate dean who advised me to stop upsetting influential people."
>
> "... advised us that we should not publish any more of our findings. We went on to present 14 abstracts, and 15 publications."
>
> "Who will blow the whistle?"

FIGURE 9.2: Peter Wilmshurst, MD, is a very forthright scientist, enthusiastic researcher, and fearless whistle-blower!

Dr. Wilmshurst himself gives a fascinating recent video presentation, "The Role of Whistleblowers in Improving the Integrity of the Evidence Base," at https://youtube.com/watch?v=Xze-yPubFIY.

What must be done to stop ethical corruption in science?

Cheating by scientists is quite easy to do, but *true researchers must always be 100 percent honest*—99 percent is not enough! Without **total honesty** by scientists and engineers, there really is no way for anyone to be certain that something is true (see chapter 1). To remove criminality from scientific research, several big changes must be made: 1) unethical conduct must be more rigorously sought, 2) research misconduct of any type by individuals or companies must be much more strongly punished, and 3) the necessity for 100 percent honesty by all researchers must be more vigorously taught to student scientists in graduate schools. There can be no valid excuses, and soft reprimands for criminal research misconduct do not do much to prevent further dishonesty. It always must be remembered, both in the United States and all over the globe, that criminality in research can directly cause damage or even death to innocent people.

Essentials to be learned in chapter 9!

1. Dishonesty by professional research scientists in academia and industry does occur and is even increasing! The answer to the question posed by the chapter title is "yes"!

2. The actual number of occurrences for research misconduct is unknown, but is probably surprisingly high!

3. Research scientists must always be 100 percent honest because research misconduct can hurt innocent people (e.g., medical patients given a new drug falsely approved as being safe), and it negates the fundamental purpose of scientific research!

4. Since the causes inducing academic and industrial researchers to become criminal cheaters are well known, dishonesty in science can and must be stopped; there can be no valid excuses for letting this outrage continue!

Reference

1. Dr.M, "Whistleblower Sues Duke University for Acquiring Research Grants via Falsified Research Publications!" Dr.M on Science, Research, & Scientists, 2016, http://dr-monsrs.net/2016/09/16/whistleblower-sues-duke-university-for-acquiring-research-grants-via-falsified-research-publications/.

> "It is simply no longer possible to believe much of the clinical research that is published …"
>
> Quoted from MARCIA ANGELL in *New York Review of Books*, January 15, 2009, http://www.nybooks.com/articles/archives/2009/jan/15/drug-companies-doctorsa-story-of-corruption/.
>
> "… much of the scientific literature, perhaps half, simply may be untrue."
>
> Quoted from RICHARD HORTON in *The Lancet* 385, April 12, 2015, 1380, http:www.thelancet.com.

FIGURE 9.3: Shocking quotations from insider scientists who are certainly in the know about dishonesty in medical research indicate that it is now frequent! These contradict the very common assumption that only a few scientists are dishonest!

CHAPTER 10

How Can You Find Out What's New in Modern Scientific Research and Technology?

Now that you are gaining a basic understanding about science, research, and scientists, you will want to get more information by yourself about novel research results, recent disputes, or the latest news that you are personally curious about. Where and how can you find such information? For nonscientists, the **Internet** is the main place to locate news about science, research, and technology. This chapter gives my advice and recommendations on using the Internet effectively, so you will not have to spend many hours wandering around and wasting time!

What if you do not have a personal computer or smartphone?

First, check whether your local public library has some available for public use. Do not hesitate to ask a librarian for help in operating their computers, searching on the Internet, or finding the info you want. Other possibilities are to see if your church or other local operations have any open for public use. Many high school and college libraries have some computers that are available for their students; you might ask if you can use one of those. If nothing else is available, consider buying one; for less than $200 you can have a computer with adequate features for your Internet use and e-mail!

A very general plan for finding info about anything totally new to you!

Before we look at specifics, here is a very **general scheme for finding info on the Internet** when you get curious about something in science and research even when you have zero background! Don't be afraid to try it!

FIGURE 10.1: *The biggest difficulty with searching on the Internet is phrasing a query so that your search engine returns precisely what you want and need, but not also zillions of other websites!*

1. First step: once you want to know more about something, look up **definitions** for two or three special terms in an ordinary dictionary.

2. Second step: determine which branches and subbranches of science are involved.

3. Third step: get a little **background** about that subbranch by reading on a **wikisite** (e.g., Wikipedia, Metapedia, and others).

4. Fourth step: decide which is the **best keyword or phrase** to seek more specific info, type that into your **Internet browser,** and then press "enter"; the search engine will return many relevant websites for your inspection and possible use.

Using websites on the Internet!

Today's Internet has oodles of information for all different levels of visitors, in the form of articles, audio files, labeled diagrams, and videos. The main problem is to find something specific just for you; to try to do that, follow the general scheme just recommended above. Probably the first screen for the dozens or hundreds of websites returned by your browser will have advertisements and the most general sources that match your query. If you can make your query more specific, then you will receive many fewer and better responses, and those will have a much greater proximity to the info you are looking for. For example, instead of entering "diseases of fruits" into your browser, try requesting "fungal infections of peaches"; as another example, instead of searching for "new kinds of batteries," try searching for "cobalt in new batteries." Be certain to inspect more than the very first website listed, but never ever try to read everything!

Because there is so very much info available at so many different websites, it is easy to get sidetracked or even lost by inspecting too much. Try to keep everything as simple and direct as possible. Decide before using your browser how much info you need and at what level it should be (e.g., simplified, popular science, a popularized account, research journal report, technical, and so on).

Where are useful websites to search for specific info and news about science, research, and scientists?

The number of websites on the Internet is enormous and increases every day! How can one decide which websites to look at first? To help you find the info you want without wasting lots of time, I recommend some good English-language websites below. Some of these offer personal subscriptions to their output (e.g., newsletters, daily additions, reports, publications, and so on), either gratis or for a fee.

For the latest news from all parts of science and technology!

If you are looking for the very latest news about anything within science and research, the following recommended websites have very good coverage. The first two are weekly international science journals that feature authoritative global news and well-written commentary, as well as research reports from scientists; both are read by many active researchers in academia and industry.

      1. *Science* (AAAS: American Association for the Advancement of Science; http://science.sciencemag.org) is based in the United States.

      2. *Nature* (http://nature.com) is headquartered in the United Kingdom.

      3. *Science News* (https://www.sciencenews.org) and *Science News for Students* (https://www.sciencenewsforstudents.org) are both issued by the Society for Science and the Public; this offer news and info very suitable for general readers about all aspects of science and modern research studies.

4. ***ScienceDaily*** (https://www.sciencedaily.com/news/) gives concise coverage about new developments and research findings in science; you can even sign up for free daily updates delivered to your e-mail inbox.

5. ***Sci-News*** (http://www.sci-news.com) gives daily coverage about research advances in all parts of science and technology; it is well-suited for general readers.

6. ***Discover Magazine: Science for the Curious*** (http://discovermagazine.com/topics/technology).

7. ***Scientific American*** (https://www.scientificamerican.com) is a classic magazine about science, mostly intended for nonscientists.

8. ***National Science Foundation—News*** (https://www.nsf.gov/news/).

Searching for news about specific fields of science!

The following sources are good places to get the latest research news for specific subdivisions of science. Don't forget that some of the general websites listed above also cover different specific areas of science.

1. TECHNOLOGY: ***Spotlight: Science News*** (https://phys.org) covers all the different areas of technology and science.

2. TECHNOLOGY: ***Popular Science—Technology*** (http://www.popsci.com/technology).

3. NANOTECHNOLOGY: ***Nanotechnology News—Latest Headlines*** (http://www.nanowerk.com/category-nanoresearch.php) features the most recent advances for research and technology in nanoscience.

4. CHEMISTRY: ***Chemistry World News*** (https://www.chemistryworld.com/news) from the Royal Society of Chemistry (UK), covers all chemistry research at industries and universities.

5. CHEMISTRY: **Science-X Chemistry News** (https://phys.org/chemistry-news/) presents specific news on all the different parts of chemistry; it is somewhat more advanced.

6. BIOSCIENCE: *SciCentral Bioscience News* (http://www.scicentral.com/B-02bios.html).

7. BIOMEDICINE: *The Scientist* (http://www.the-scientist.com/?home.home) features new info about biomedical scientists and science.

8. HEALTH AND MEDICINE: *NIH News in Health* (https://newsinhealth.nih.gov) presents a variety of recent info about health advances, medical research, and objectives for personal health.

9. HEALTH AND MEDICINE AT THE NIH: *NIH Research Matters* (https://www.nih.gov/news-events/nih-research-matters) gives weekly accounts of research progress at the US National Institutes of Health.

10. MEDICINE: *SciCentral Health Science News* (http://www.scicentral.com/H-02heal.html).

11. MATERIALS SCIENCE AND ENGINEERING: *Engineering Materials News* (http://www.materialsforengineering.co.uk/engineering-materials-news/) covers the latest progress in developing diverse improved materials for practical or research uses.

12. SPACE SCIENCE: **NASA** (https://www.nasa.gov) lets visitors observe the many different activities for research and exploration at the US National Aeronautics and Space Administration. There is very much to see, and you can easily and enjoyably spend all day exploring here!

For info regarding honored research scientists!

The following three websites can be explored for many interesting hours both by raw beginners or by expert scientists!

1. "Listing of Nobel Prizes and Laureates" (https://www.nobelprize.org/nobel_prizes/).

2. "Kavli Prize Laureates" (http://www.kavliprize.org/prize-landing-laureates).

3. "Breakthrough Prizes" (https://breakthroughprize.org/).

For info regarding specific scientists!

Most faculty scientists in academia maintain personal websites/blogs on the Internet, dealing with their research projects and showing some pretty data from their experiments. Industrial scientists have websites listed under the name of their commercial employer. If you are curious about a recent research report, visit the website of the chief or corresponding author.

Other sources in or out of the Internet for information about science!

Many **professional science societies** have special educational features on their websites aimed at science students or adult nonscientists. Ask your browser to list websites for specific science societies.

Certain **people** might be very helpful for finding the info you are seeking. These include authors of science books and research publications in science journals, doctoral scientists working at universities or industrial R&D centers, or workers associated with various science-related associations, businesses, clubs, or stores; don't forget to ask your past or present science teachers for some recommendations.

If your searches on the Internet have lead nowhere, and you are getting desperate to find something, you can ask **your children** (at any level of school!) if they have covered that topic in their science classes. If so, take a look at their textbook or course instruction materials; those will give you a useful background for your subsequent searches on the Internet.

Lastly, I will recommend here only one special website that covers many different topics and issues in science, because I particularly like this one: "Dr.M on Science, Research, and Scientists" (http://dr-monsrs.net)!

A specific example of finding info: "What are gravitational waves?"

Announcements about the winners of big prizes for research excellence usually raise many questions about what their investigations discovered, why their findings are very significant, and how their discoveries impact problems in everyday life. A typical real example arose with the recent 2016 Kavli Prize awards in **astrophysics**, where decades of research work finally discovered solid evidence that **gravitational waves** really do exist in nature; hence, many people with curiosity about those waves must search to find enough explanation so they can understand this prize-winning research accomplishment.

Even most doctoral scientists in other fields of science (e.g., me!) had never heard of gravitational waves before the announcement of the new Kavli Prize winners! They had to learn more about them, just like you must do, in order to understand that award. Both you and all those scientists can approach this task by using the Internet. In learning about gravitational waves, always recall that you are *not* looking to become a world expert, but only want to learn enough so that you can have a basic understanding!

Before using a browser for this example, it always is useful to first get definitions of a few special terms involved (i.e., astrophysics, gravity, waves), and then to acquire some background. You will find that astrophysics is a major branch of astronomy (i.e., physical science > astronomy > astrophysics) that primarily studies all the different comets, exoplanets, galaxies, stars, and other components of the universe.

Now you will have enough background to be able to understand the descriptions and explanations given about gravitational waves on wikisites (physical science > astronomy > astrophysics > gravitational waves). Look at explanations about what they are, how they arise, how they are detected, and how they can be measured. Although the theoretical concept of gravitational waves arose decades ago, scientific research could not prove they existed in nature until very recently; this very significant breakthrough justifies awarding the 2016 Kavli Prize and the 2017 Nobel Prize to researchers achieving this advance. With that, you should now be able to comprehend the concept of gravitational waves.

Bingo, you have just successfully educated yourself!

You also can do the same for any other subjects that have special interest for you! Write down in a notebook some successful key findings coming from each of your searches on the Internet. Also, insert printouts of any particularly good labeled diagrams. Limit yourself to just one or two pages for each search; that will force you to be concise! By keeping this notebook, you will not have to try to remember all that you have found, and your notes will be very helpful for Internet searches several years later. Good luck with all your further efforts to learn and understand more about science, research, and scientists through using the Internet!!

Essentials to be learned in chapter 10!

1. The Internet provides general and detailed information about very many topics in scientific research; a general sequence is recommended for looking up info on the Internet, even if you start from total ignorance.

2. Many other useful sources of information about science, research, and scientists that do *not* involve the Internet also are available: living scientists, libraries and librarians, professors and science teachers, certain topical magazines, and so on.

3. A number of good websites in English that deal with science, research, and scientists are recommended by Dr.M!

FIGURE 10.2: *The Internet is amazing because it is useful for so many different activities; everyone should learn how to use it, and no one is too senior to be unable to enjoy using the Internet!*

CHAPTER 11

What If You Want to Do Some Volunteer Work with Scientists on a Research Project?

If you become enthusiastic and really interested in science, how can you get to work on a research project with scientists? To actually go beyond just reading about research, you *first* must recognize that doing good experiments is not so easy, and is much more than having creative ideas or working with computers. It requires a very strong personal commitment, because it definitely is not a casual pursuit! *Second*, anyone wanting to get their hands wet with research must accept that they need to acquire a lot more learning and experience than whatever they already have. Research scientists have had advanced courses, much practical experience, 1:1 instruction about instrument operation and research methods, technical training, and so on; chances are you do not already have those.

This chapter takes a look at how you might participate in scientific research without first getting a PhD!

Some very necessary questions if you wish to participate in research now!

Your biggest and most important question will be, "What exactly is your number one reason for now wanting to do some research work?" Many reasons are very likely to indicate only a shallow mentality and limited understanding about what is involved. Other important questions include the following: Which part of science interests you the most, and why? Have you already done any preparation for research work? Have you written a term paper in school about something in science? How does doing some research now fit into your plans for the future? How long would you want to work in a research lab if full-time cleaning and inventory work was your assigned job? All these questions are only intended to force you to be totally realistic with your expectations!

FIGURE 11.1: *Volunteering to work on scientific research usually is difficult to do; however, anyone can be totally interested in science, research, and scientists!*

Could you work on scientific research as a volunteer?

The best answer to this question is "possibly, but..."! That short answer can be completed by adding that volunteer positions are not numerous, you probably will not get paid any salary, and you might find your assigned work to be boring! What if you are accepted to work in a lab as a volunteer, but are assigned to be an extra pair of hands sorting many thousands of minute specimens into a dozen defined groups? Or, what if you were assigned to work only on some other very routine type of task? That could happen very easily! *Nobody can work on research experiments as a volunteer until they have a certain level of education and special training*!

If you wish to try research work in order to see whether it might suit you as a later career, then something might become available provided you can find the right person to let you join his or her project. Ask yourself, "Exactly how would my volunteer efforts help the lab I will work in?" If you are only looking for having some fun during a summer, then your motivation is likely to be seen as insufficient.

Are there organized programs for volunteers to work on research? Yes!

The better current possibilities for you to volunteer for working on a research project are available to almost anyone! You can apply to one of the organizations dedicated to involving ordinary people with science and research activities. **Citizen Science** (http://citizenscience.org), **SciStarter** (https://scistarter.com/citizenscience.html), and several others are available on the Internet. These are large organized programs for persons having limited knowledge about science to volunteer as active participants in selected research programs and projects. Their websites contain details and descriptions of research projects currently needing more volunteers!

In addition, some schools and large industries have organized programs for limited-term work in a research lab. Definitely start investigating a year before you want to begin. If you are still in school, your quest is somewhat easier. *For students in secondary schools,* ask one of your science teachers if there are any research activities available; for example, you might want to start work on a new project for the Science Talent Search (https://student.societyforscience.org/regeneron-sts), an annual competition for research done by high schoolers. Or, your science teachers might know a professional researcher he or she could introduce you to.

Students who currently are enrolled at an undergraduate college might want to work in a research course leading to a senior thesis. Yet another possibility is to find out if your college has any special laboratory research programs available for their enrolled students. Any of those opportunities will help you get a real taste of lab-based research work!

Some brief advice for potential volunteers!

Except for the different possibilities described above, finding a volunteer position with an ongoing real research project is generally not easy. This unfortunate situation is changing since there is more and more recognition that both younger and older persons will increase their interest and support for science, research, and scientists if they can become personally involved and actively participate. On the other hand, the door is wide open for anyone to become interested and enthusiastic about science and research via the Internet (see chapter 10).

Actually, doing hands-on work in a research lab is *extremely different* from reading about science! Whether you wind up getting enjoyment by reading and watching videos about science, research, and modern technology or getting satisfaction by actually helping produce new knowledge in a research laboratory, you must make your own decision about what is best for you. Either choice will be quite OK!

My guess is that most readers of this book will simply want to learn to understand more about science and research, and will *not* be determined to personally conduct research experiments. It definitely is *not necessary* to do lab work in order to be very, very enthusiastic about science and completely fascinated by scientific research!

Essentials to be learned in chapter 11!

1. Volunteer positions working on lab research are not many, and finding one is not easy!

2. Some organized programs for ordinary adults to actively participate in scientific research projects as volunteers are available on the Internet.

3. If you are still a young student and want to evaluate whether your capabilities, interests, and personality are suitable for a later career with scientific research, then you should find out whether your current school or a nearby institution has any available programs with hands-on activities.

FIGURE 11.2: Anyone can be deeply interested and enthusiastic about science and research without actually conducting experiments and being a scientist! Professional scientists will generally welcome your personal interest in their research investigations!

CHAPTER 12

What Can Science and Research Do for You?

Has your life already started to become more interesting now that you are learning to understand more about science and research? Although few people realize it, everyone's daily life directly depends on scientific research and engineering developments. The food you eat for breakfast, the clothes you wear, the gasoline you put into your car, the water you drink, the garbage you dispose of, and the technology that allows you to watch live TV all directly depend upon the research and development work of scientists and engineers. That is a fact, and it is true whether you are an aerobics instructor, airline stewardess, a banker, construction worker, farmer, parent, salesperson, secretary, teacher, waitress in a restaurant, or the CEO of a giant international corporation!

Research is the means whereby science makes discoveries, advances technology, and solves or alleviates many practical problems! This chapter will show you how to further develop your understanding about how research nowadays is important for everyone. It presents several examples of problematic everyday situations that you might encounter, and then shows how to deal with them better by learning more about science.

Example #1: Every winter you get several colds despite all your efforts to only eat healthy food and avoid being near anyone who is sneezing or coughing!

Advertising	Frogs	Palace
Airport	Garbage	Paper
Art	Hurricanes	Pasta
Balloons	Intestines	Payday
Beach	Lies	Peaches
Chorus		Printer
Colors		Seattle
Coughing		Snow
Death	SCIENCE	Sunlight
Dragonfly		The Kinks
Drama		Toys
Dust		Travel
E-mail	Money	Trees
Fat-free	Monitor	Whistles
Flavors	Music	Wind
Fog	Noise	Windows
Fragrance	Owls	X-rays

FIGURE 12.1: *Even if almost nobody realizes it, science and research are everywhere and are important to everyone each and every day! That even includes you! How truly amazing!*

Common colds are caused by infection with a "cold virus." Do you know what a virus is and how they infect us? Most antibiotics kill bacteria, but do not have a lethal effect against viruses; therefore, taking common antibiotics for your cold is pointless and won't help you. Options for treatment to relieve symptoms are readily available (thank goodness!), including some from folklore. What do you do when you get a fever? Fever is nature's way of fighting infections by microbes (viruses and bacteria), yet most people try to stop or reduce fevers! Fever is good and will help you fight your cold, so long as it does not rise to dangerous levels! Of course, any fever in infants needs the immediate care of a pediatrician, and any high or ongoing fever in adults should have its cause determined by a physician.

Begin supportive measures: take a day or two off from work and get some extra sleep, keep warm, eat chicken noodle soup and fresh oranges, and drink lots of fluids. Do you know how these help your body and cells fight the effects of infection by a cold virus? While resting, you can do some reading on the Internet (see chapter 10) about antibiotics, colds, fever, viruses, and so on; then you will become more knowledgeable about what a common cold is, how your viral infection began and spreads, how your body fights infections with cold viruses, and how restorative measures can help you to recover.

After the respiratory symptoms of your cold decrease and vanish, this new knowledge will prepare you to better understand why research in medical science still has *not yet* been able to find a general cure for the common cold. Viruses are hard to kill because they are a moving target! After your immune system forms lethal antibodies against a cold virus during a previous infection, that old virus can undergo mutations. Mutations are genetic changes that occur randomly and often; when this causes changes in the protein structure of a cold virus, your antivirus antibodies formed against the previous infective cold virus might no longer bind and neutralize the mutated cold virus. That means the old virus causing your current cold is actually "new" to your immune system, and thus you can easily get yet another cold from it! Yes, viruses are indeed clever and very tricky! Also, you will be amazed if you look up images of virus structure on the Internet!

Example #2: Your young daughter continues to get poor grades in her science classes at school, and you are too ignorant to be able to help her!

To assist your daughter to learn, you also must learn more! You can shorten the time needed to do that by focusing only on the next topics scheduled to be studied in her science class. Read about the next topic in your daughter's textbook and teaching materials; if this is too difficult to comprehend on your first effort, then simply read or watch everything again for a second or third time! You can do it!

Encourage your daughter to learn better by discussing her latest science lesson. Using a "question-and-answer" approach, you will first ask her some questions, and after she responds then she should present a few questions for you to answer. Try to locate some good instructional videos on the Internet (see chapter 10) that are specifically made for youngsters to learn about many topics in science; then, view several of these together and discuss their educational messages with her.

Youngsters often learn better with situations directly involving themselves instead of trying to learn from dry material in a textbook. Show your daughter something new or useful about science that she can directly relate to in her own world. Whether she has interests in aerobics, art, computers, horses, organic food, swimming, or the nature of glass is irrelevant; using the above approaches, you can help her learn about science with any of those! Not only will your daughter's grades improve, but you will get to know and understand much more about science yourself!

Example #3: Your old car still works well, but its body is getting more rusted! What can you do to make it stay intact for a few more years since you can't afford to buy a new one?

To answer this query, you need to learn more about: 1) what rust is, 2) which weather conditions stimulate rusting and which inhibit rust formation, 3) what commercial products are available to try to stop further rusting, and 4) how do those work? You will find answers and info using any browser to search on the Internet (see chapter 10). By acquiring more knowledge and understanding, you will increase your ability to take effective actions to either stop or at least inhibit further rusting; your old car then should continue to remain useful for several more years.

You might be very surprised to realize that reading your new info about rust and rusting is actually learning more about chemistry! Rusting is a chemical reaction between iron and oxygen. There are many equations to describe it, but you do *not* need to learn any equations in order to increase your general understanding! What you are after is to find out the basic causes and effects of rusting. Once you understand those, you will be able to take effective steps so further rusting will be delayed or even prevented from taking place.

How is learning more about science and research valuable for handling life's everyday problems?

The three examples just presented all use the *same general sequence of steps* for learning to solve some of your problems: 1) First, recognize that a practical problem is present and be as specific as possible. 2) Then, use the Internet to gain some background about the possible causes and effects involved with your problem area. 3) Next, use your new knowledge to decide about the most likely possibility for taking effective counter-measures. 4) Last, try out the easiest or best possibility yourself and see what kind of result you obtain; if your first attempt does not work, then try something else. Good luck!

Knowledge about science by itself is simply there, but it magically transforms into something useful when it leads you to deal more effectively with common problems. For that purpose, getting new knowledge and

developing a more detailed understanding about something in science stops being dry and boring, starts becoming helpful, and even can be fun!

How do science and research make life much more interesting?

Scientific research, engineering developments, and new technology can be applied to anything! That includes making new batteries or new babies, treating and even curing cancers, evaluating the possibility of capturing comets or meteors in space for bringing them back to Earth where their valuable mineral content can be extracted, finding ways to generate electricity that are cleaner and less expensive, treating birth defects by manipulating genes or stem-cells, regenerating body parts in senior citizens, and developing household robots and driverless electric autos.

Even designing and producing better shoes (i.e., less expensive, longer lasting, more comfortable, warmer or cooler, and so on) can involve the work of scientists and engineers! However, research and development can also produce new ways to kill more people in wars, so be sure to recognize that scientific research is ultimately a two-way street!

You can now go ahead and put everything in this chapter to a personal test! Pick whatever subject or problem is most personally important for you right this minute and look up what are the latest relevant research findings and developments by searching on the Internet (see chapter 10). After reading several descriptions or viewing a video about the newest findings, you will soon become more aware of how scientific research has direct effects upon you and everyone else. It always is quite interesting, so when your curiosity is aroused, you will acquire many new ideas and fascinating stories to discuss with your friends!

Essentials to be learned in chapter 12!

1. Scientific research discoveries and engineering developments often have practical uses; learning about them will give you a better ability to deal with many practical problems or difficulties popping up in your own everyday life.

2. Having a greater understanding about science and research will also make your life much more interesting! Try it and you will be amazed to see it happen!

SCIENTIFIC RESEARCH HELPS EVERYONE EVERYDAY!

Cell Biology Research → producing stem cells
Chemical Research → new gasoline
Computer Research → cheaper, smaller PCs
Drug Research → treatment for orphan diseases
Engineering Research → better desalination
Nutrition Research → much safer food
Materials Science Research → better batteries
Medical Research → increased curing of cancer
Polymer Research → better contact lenses
Software Research → secure online banking
Systems Research → driverless vehicles

FIGURE 12.2: *If only everyone could realize just how much scientific research is helping you and everybody else right now and every single day!*

INDEX

name or topic	page numbers
academia	11
Angell, Marcia	73,78
applied research and applied science	5,33
assays	5
basic research and basic science	5,33
Big Science	49
biology and medicine (biomedicine)	3
Breakthrough Prize	57
cheating at research	71
chemistry	3
Citizen Science	91
conduct of research investigations	18
coworkers in research laboratories	27
cost of scientific research	37,38,41
criminality in science and research	71,74,75
dishonesty by scientists has bad effects	74
doctoral thesis	13
engineers	5,26
evidence for The Truth	4,7,8

experiments	4,6,18
faculty scientists	31,33,34
Farquhar, Marilyn G.	67,68
Fischer, Artur	25,61,63,64
fun in research	19,20
graduate school	13
graduate students in science	13,26
groupings for conduct of research	45
Horton, Richard	74,78
how are new scientists produced?	13
how do we know what is true or false?	8
industrial scientists	14,31--35
individualism	48,66
Internet	79,81
browsers	81
wikisites	81
inventors	23,25
Kavli, Fred	56
Kavli Prize	56
lab technicians (research techs)	26
laboratories (labs)	6,45,46
laboratory research	27,28,46
Land, Edwin H.	64,65

Little Science 45,49
Mullis, Kary B. 66,67
Nobel, Alfred 53,54
Nobel Prize 53,55,56
Obokata, Haruko 73
patents 19,33
physics (physical science) 3
PhD (doctor of philosophy degree) 13
postdocs (postdoctoral research associates) 13,27
principal investigator (PI) 23
protocols 5
R&D (research and development) 5,6
research (scientific research) 4,5,6
    breakthroughs 28
    data 4
    doctoral thesis 13
    dollars for research expenses 37,38,39,41
    experiments 4,5,18
    factories 47
    freedom 33,34
    giant research groups 46,47
    grants 40
    industrial model for research 50

experiments	4,6,18
faculty scientists	31,33,34
Farquhar, Marilyn G.	67,68
Fischer, Artur	25,61,63,64
fun in research	19,20
graduate school	13
graduate students in science	13,26
groupings for conduct of research	45
Horton, Richard	74,78
how are new scientists produced?	13
how do we know what is true or false?	8
industrial scientists	14,31--35
individualism	48,66
Internet	79,81
browsers	81
wikisites	81
inventors	23,25
Kavli, Fred	56
Kavli Prize	56
lab technicians (research techs)	26
laboratories (labs)	6,45,46
laboratory research	27,28,46
Land, Edwin H.	64,65

Little Science	45,49
Mullis, Kary B.	66,67
Nobel, Alfred	53,54
Nobel Prize	53,55,56
Obokata, Haruko	73
patents	19,33
physics (physical science)	3
PhD (doctor of philosophy degree)	13
postdocs (postdoctoral research associates)	13,27
principal investigator (PI)	23
protocols	5
R&D (research and development)	5,6
research (scientific research)	4,5,6
breakthroughs	28
data	4
doctoral thesis	13
dollars for research expenses	37,38,39,41
experiments	4,5,18
factories	47
freedom	33,34
giant research groups	46,47
grants	40
industrial model for research	50

	large research groups	47
	managers	25
	misconduct	73
	organization	31,45
	progress in academia	28,35
	progress in industries	35
	projects	6,19
	questions	5,7
	small research groups	45
	staffing of laboratories	23--27,33
	subjects, specimens	7
	technicians	26
	thesis research project	13
salaries for doctoral researchers		39
scientists		6,7,11,14
	daily activities in academia	14--19
	education and training	11,13
	employment	14,46
	faculty scientists	31,33,34
	individuals as scientists	61--68
	industrial scientists	31,41
	R&D work in industries	31,41
	salaries	39

scientists as managers	25
soft- or hard-money salaries	39
teamwork for research	28,50
SciStarter	91
the laboratory family	28
thesis adviser	13
total honesty	77
undergraduate students	26,92
value of science and research	95,96,101
volunteering for research work with scientists	89
what is scientific research?	4,5,6
what is science?	1,2,3
whistle-blowers	75
who pays for scientific research?	37,41
Wilmshurst, Peter	75,76

www.ingramcontent.com/pod-product-compliance
Lightning Source LLC
Chambersburg PA
CBHW040220220526
45473CB00001B/55